TWENTIETH-CENTURY SHORE-STATION WHALING IN NEWFOUNDLAND AND LABRADOR

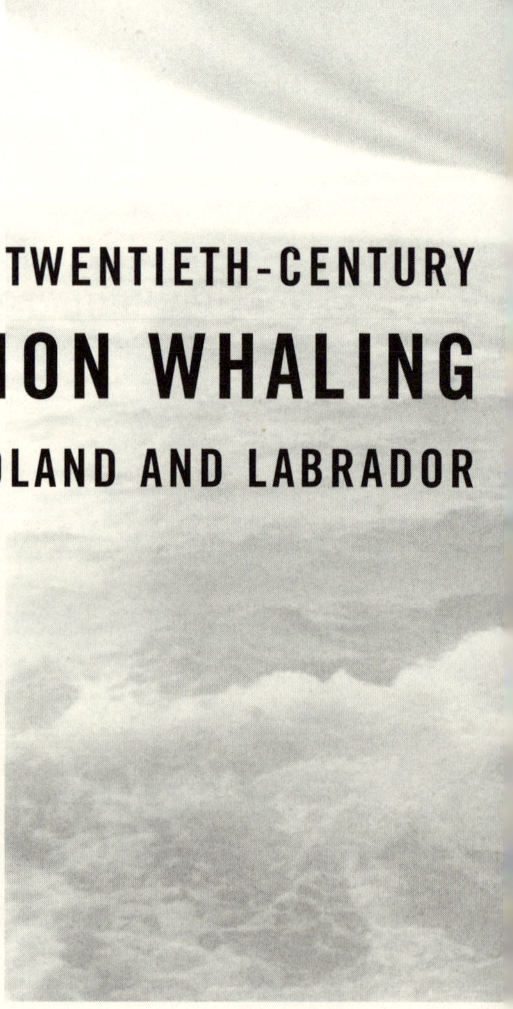

TWENTIETH-CENTURY

SHORE-STATION WHALING

IN NEWFOUNDLAND AND LABRADOR

McGILL-QUEEN'S UNIVERSITY PRESS Montreal & Kingston · London · Ithaca

Anthony B. Dickinson and Chesley W. Sanger

ISBN 0-7735-2881-4

Legal deposit second quarter 2005
Bibliothèque nationale du Québec

Printed in Canada on acid-free paper that is 100% ancient forest free (100% post-consumer recycled), processed chlorine free.

This book has been published with the help of a grant from the Canadian Federation for the Humanities and Social Sciences, through the Aid to Scholarly Publications Programme, using funds provided by the Social Sciences and humanities Research Council of Canada.

McGill-Queen's University Press acknowledges the support of the Canada Council for the Arts for our publishing program. We also acknowledge the financial support of the Government of Canada through the Book Publishing Industry Development Program (BPIDP) for our publishing activities.

Library and Archives Canada Cataloguing in Publication

Dickinson, Anthony Bertram
Twentieth-century shore-station whaling in Newfoundland and Labrador /
Anthony B. Dickinson and Chesley W. Sanger.

Includes bibliographical references and index.
ISBN 0-7735-2881-4

1. Whaling–Newfoundland and Labrador–History–20th century.
I. Sanger, Chesley W. II. Title.

SH383.5.C3D52 2005 338.3'7295'097180904 C2004-907023-1

Design and typesetting by studio oneonone in Adobe Garamond 11/12.3

For Max and Ethan Dickinson, and Anthony, Emma, and Samuel Clark

For Andrew Sanger and six of his great-grandchildren:
Kielyn, Cassidy, Drew, Rebecca, Tyler, and Leah

CONTENTS

FIGURES AND TABLES

FIGURES

TABLES

ILLUSTRATIONS

PREFACE

The economy of Newfoundland and Labrador has historically depended on the exploitation of natural resources, especially cod and seals. The operations and socioeconomic impacts of these fisheries are well understood. What has not been well documented is the local "modern" shore-station whaling industry in the years between 1898 and 1972. Although whaling did not have the same overall economic impact as cod fishing and sealing, it provided important employment for residents of communities adjacent to the stations. The narrative that follows examines the conduct and impact of this small but important phase of marine resources utilization in Newfoundland and Labrador, and of global whaling.

Commercial whaling was often highly cyclical, with each cycle proceeding through stock discovery, exploitation, expansion of the industry and increased competition, and resulting stock depletion. The introduction of more sophisticated technologies and techniques reduced stocks further. Whaling became unprofitable, and operations closed. The modern whaling industry of Newfoundland and Labrador followed the pattern.

This study examines the seasonality of operations, changing spatial relationships, and environmental and biological factors that influenced the origin, rise, and decline of each contiguous cycle of the local industry. Information was collected from archives in Newfoundland, Canada, the United States, Scotland, England, and Norway. The principal primary sources included government papers, reports, legislation, and statistical summaries, whaling company documents and data, public and private petitions, oil and whalebone manufacturer lists, pamphlets, diaries, letters, and telegrams. Information was usually sporadic, incomplete, and in the case of catch data, variable in different sources. Newspapers were primarily used to develop the framework of the study and the industry profile. They provided statistics on oil and bone production and prices, vessels, masters, companies, ports, hunting grounds and catches, first-hand accounts and explanations of decision-making, viewpoints, and the hopes and fears of active participants. The substantial body of secondary whaling literature yielded little directly on the Newfoundland and Labrador industry and was mostly used to place the local trade in a global context.

Four major depositories of primary information were examined outside Newfoundland and Labrador. The Smithsonian Institution contains papers and photographs of whaling at Snook's Arm and Balaena obtained by personnel visiting at the beginning of the twentieth century. The Provincial Archives of British Columbia provided documents giving comparative insight into the rise and fall of Canadian West Coast whaling, and its relationship to the Newfoundland industry before World War I. The papers of Christian Salvesen Co. Ltd of Leith, Scotland, held in the University of Edinburgh library, document the company's participation in Newfoundland and Labrador whaling during the 1930s and World War II. The only comprehensive set of correspondence and operational records found to date of an early Newfoundland whaling station (Aquaforte) is in the Chr. Christensens Hvalfangstmuseum at Sandefjord, Norway. These were previously in the possession of the Kiserud family of Oslo, descendants of those who established the factory.

The research for this study was supported by grants from the Social Sciences and Humanities Research Council (SSHRC) of Canada, and the Vice-President's Fund, and the Institute of Social and Economic Research (ISER), Memorial University of Newfoundland. Publication was supported by the Social Science Federation of Canada using funds from the SSHRC and Office of Research, Memorial University of Newfoundland.

We are indebted to Philip Cercone, executive director and chief editor at McGill-Queen's University Press, for taking on the publication of this work, to Joan McGilvray, coordinating editor, for overseeing the publication process, and to Maureen Garvie for her skilful editing, and to the typesetters for their excellent work. We also thank the readers of McGill-Queen's University Press and the Aid to Scholarly Publications Programme (SSHRC) for their helpful comments.

Memorial University of Newfoundland also contributed student assistance to the research through its Summer Career Experience Program. Its Cartography Laboratory (Department of Geography) helped prepare maps and figures, and for this we thank C. Conway, D. Curtis, F. McDonald, G. McManus, and D. Mercer. We are also grateful to Anne McLeod, Anigraph Productions Ltd, St John's, for some illustrations, to Prof. Lewis Fischer, Department of History, and Dr Gordon Handcock, Department of Geography, for their insightful and ongoing comments, and to Dr James Mead, Curator of Marine Mammals of the Smithsonian Institution, for access to the collections of Dr F.W. True. Børre and Egil Kiserud and their families, including their mother,

Marie (born in Newfoundland in 1904), welcomed us warmly into their homes and translated the Norwegian material on Aquaforte. Preparation and editing relied heavily on Sharon Wall, Martha Sanger, Ann Sanger, Linda Jackman, Marilyn Hicks, Peggy Parsons, and Eileen Ryan. Edwina Ryan and Kevin Hicks assisted with reproductions. Our thanks to them all.

Others from Newfoundland who provided information and photographs included Annette Anthony, librarian, Dept. of Fisheries and Oceans; Sandra Ronague, Still and Moving Images PANL; Thomas Burry, Carole-Anne Coffey, Marguerite Coffey, Edward Cokes, John C. Crosbie, Dennis Flynn, Vaughn George, Maggie Hennessey, the family of the late Thomas E. Hennessey, Bruce Hodder, Robert Kelly, George Lilly, Augustus McGrath, Bruce Melvin, Ada Nemec, Jacqueline Perry (DFO), Jolliffe Quinton, Lee G. Shinkle, Robert Thorne, and Jean Wiseman. From elsewhere in Canada we acknowledge Graeme Balcom of West Vancouver, B.C., Mr and Mrs G.B. Elliott, Joan Goddard and Olive Miles of Victoria, B.C., Wyb Hoek of Mont-Joli, Quebec, and Iain Reid of Ottawa. In the U.S.A., we thank Ralph E. Eshelman, Solomons, Maryland; Richard Street, Bay City (Gray's Harbour), Wa.; Dale Vinnedge, Freemont, Calif.; and Nicholas Whitman, Williamstown, Mass. In Germany we acknowledge Klaus Barthelmess, Cologne, and Axel Kuehn, Karlsruhe, and from Norway, Hakon Sorvald, Oslo.

We are also grateful to the staff of the Aberdeen Public Library, Wa.; American Museum of Natural History Library, New York; British Columbia Provincial Archives, Victoria; Centre for Newfoundland Studies, Queen Elizabeth II Library, Memorial University of Newfoundland, St John's; City of Vancouver Archives; Dildo and Area Interpretation Centre, Dildo, Newfoundland; International Whaling Commission, Cambridge, England; Kendall Institute of the New Bedford Whaling Museum, Mass.; Chr. Christensens Hvalfangstmuseum, Sandefjord; the library of the Agassiz Museum of Comparative Zoology, Harvard University, Mass.; the library of the Department of Fisheries and Oceans, St John's; Maritime Museum, Bergen; Maritime Museum of British Columbia, Victoria; Nanaimo Centennial Museum and Archives; National Museum of Natural History, Smithsonian Institution, Washington, D.C.; Norsk Sjøfartsmuseum, Oslo; Norsk Teknisk Museum, Oslo; Norwegian Museum of Science and Technology, Oslo; Provincial Archives of Newfoundland and Labrador, St John's; Provincial Reference and Resource Library, St John's; Public Archives of Canada, Ottawa; University of British Columbia Library, Special Collections Division, Vancouver; Vancouver Maritime Museum; the Vestfjold Fylkesmuseum, Tønsberg; and the Whaling and Sealing Museum, South Dildo, Newfoundland.

Portions of this book have appeared in preliminary form in other publications or have been presented at symposia. These are referenced in the chapter notes and bibliography.

TWENTIETH-CENTURY SHORE-STATION WHALING IN NEWFOUNDLAND AND LABRADOR

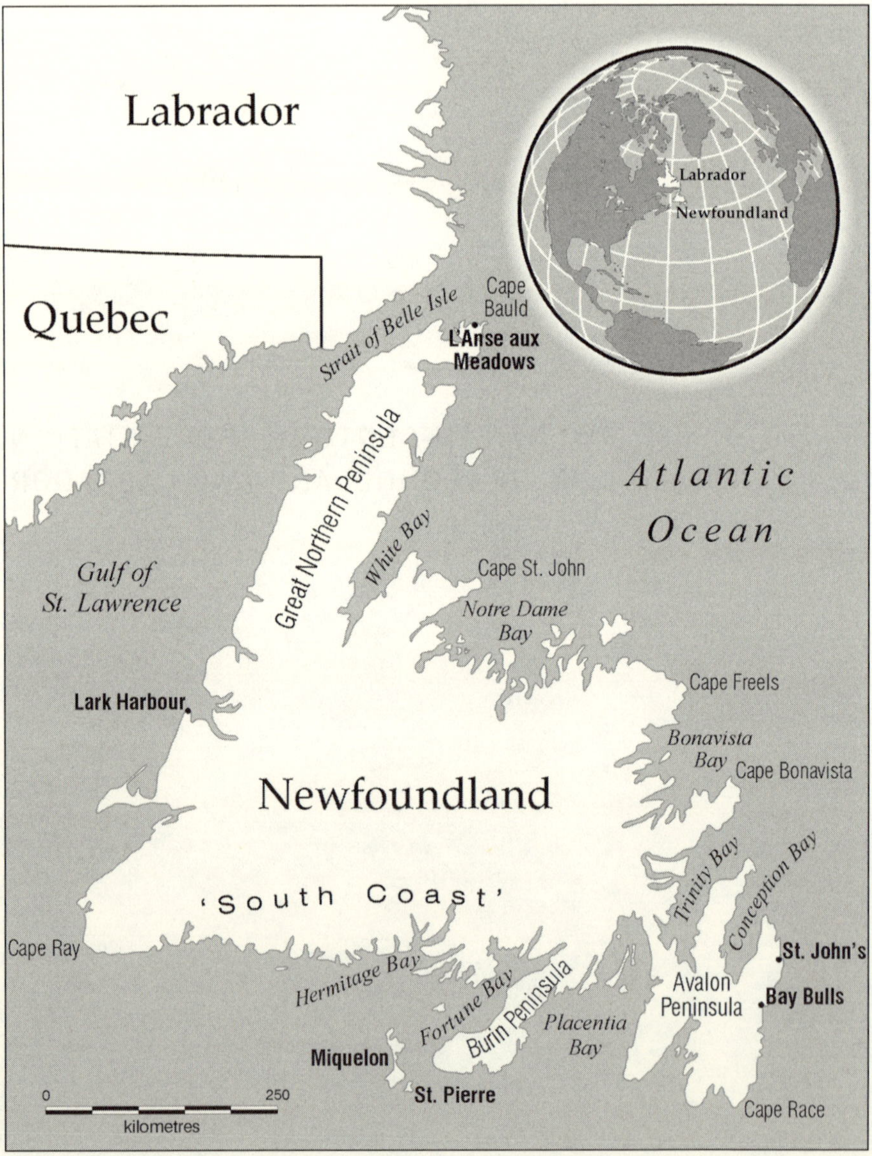

Fig. 1.1 Location map

CHAPTER 1

The Setting

NEWFOUNDLAND AND LABRADOR

Newfoundland and Labrador,[1] Canada's most easterly province, is located adjacent to the biologically productive waters of the Grand Banks (fig. 1.1). The first inhabitants were the Maritime Archaic Indians who arrived in Labrador around 7,000 B.C. as the last ice-cap retreated. They moved into insular Newfoundland by about 3,000 B.C., and survived by gathering, fishing, and hunting, along the littoral. This culture was gradually replaced in Labrador from about 2,000 B.C. by the Palaeo-Eskimo whose ancestors originated in Alaska and moved eastward across the Canadian Arctic. The later Dorset Palaeo-Eskimo dispersed widely around the coast of Newfoundland from approximately 500 B.C. Artifacts indicate that they used marine mammal oil for cooking and heating. The first local indigenous people to persistently hunt whales were the Inuit, who arrived in coastal Labrador around 1,500 A.D. from Alaska. More recently, Beothuk Indians lived on the island of Newfoundland until 1829 when the last known survivor, a woman named Shanadithit, died in St John's.[2]

The first European presence occurred around 1,000 A.D. when Norse from west Greenland explored the coast of Labrador and established a short-lived settlement[3] at the tip of the Great Northern Peninsula of Newfoundland.[4] The colonists may have been able to survive for a time by catching whales from those stocks that annually migrated through the Strait of Belle Isle and later supported commercial fisheries.[5] Nearly five centuries elapsed before Europeans returned.[6] In 1497 John Cabot discovered a "new isle"[7] – Newfoundland – with an abundance of cod while attempting a voyage to the Orient. Other explorers followed,[8] and news of the prolific fish stocks spread rapidly throughout maritime Europe.

Within a few years fishermen began sailing from Portugal, Spain, and France to this "New-Founde-Land." By the end of the sixteenth century the Iberian fishers were being supplanted by those from southern and western England. During much of the next three centuries they shared and fought with the French over this northwest Atlantic migratory fishery.[9] The industry gradually changed from seasonal to residential, and cod became and remained the economic staple of Newfoundland and Labrador.[10] This continued until 1992 when the Government of Canada placed a moratorium on commercial cod fishing along the northeast coast and Labrador because of depleted stocks.[11] Catches had declined significantly from the 1970s, mainly because of increasingly effective catching technologies and overfishing by foreign and domestic vessels. Diversification into unutilized or underutilized species such as crab and shrimp increased the value of the industry, although the earnings were shared among a significantly smaller number of people.[12]

Permanent settlement occurred largely to exploit cod, and communities were thus located close to prolific fishing grounds.[13] Hundreds of isolated fishing stations, or outports, often only connected by small boats, developed along the coast.[14] The harsh realities of the environment (infertile land, rugged topography, sea ice, and rigorous winters) often restricted travel between them.[15] The household was the fundamental economic unit, producing salt cod, initially with the help of indentured servants. Most families managed to sustain themselves, albeit precariously, by marketing their fish and by subsistence activities such as keeping livestock, growing vegetables, and harvesting seals, sea birds, and game (fig. 1.2).[16]

From the mid-seventeenth century, the majority of English migrants and settlers came from southwest England, especially Devon, Dorset, western Hampshire, and southern Somerset (fig. 1.3). Migration patterns were largely determined by the location of their respective trading establishments in Newfoundland. Those from Dorset, for example, dominated the northeast coast where trade and commerce were monopolized by merchants from Poole through their establishments in places such as Brigus, Bay Roberts, Carbonear, Trinity, Bonavista, Fogo, and Twillingate.[17] After about 1720 the number of Irish entering the Newfoundland fishery grew significantly as the English vessels increasingly called at ports such as Waterford, Wexford, and Cork for supplies and labour (men and boys) on their outward voyages in the spring.[18]

The premises of the English firms became centres from which food, clothing, and fishing supplies were distributed to nearby outports.[19] Although fishing families were nominally independent, the "truck," or credit system kept most indebted to the merchants who outfitted them with gear and supplies on credit and took their processed fish and oil in payment.[20] (This would persist to some extent until the salt fish industry collapsed in the 1950s.)

In the development of permanent settlement in Newfoundland, the French Revolutionary and Napoleonic Wars were an important watershed. The seasonal flow of fishermen from England and Ireland was greatly reduced, and merchants encouraged more of their agents and employees to remain on the

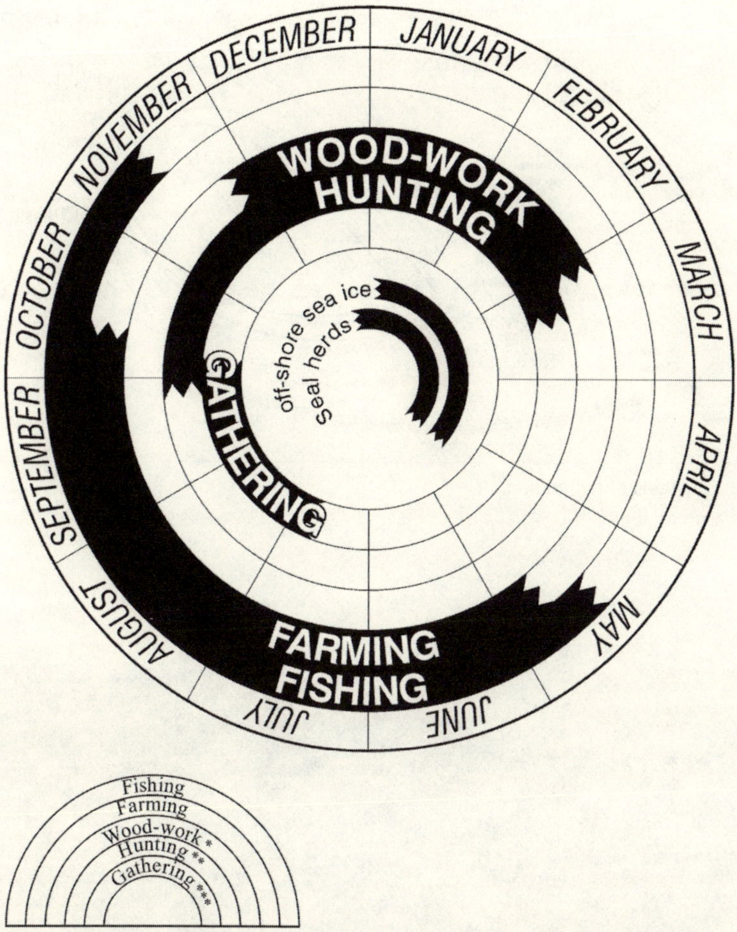

* Wood-work; Fire-wood, fencing materials, saw-logs, and other building materials
** Hunting; Birds, terrestrial mammals
*** Gathering; Wild fruits and berries
NOTE: All boundaries are approximations and represent decreased activities

Fig. 1.2 Rural northeastern Newfoundland, activity cycle of a subsistence household.
Source: Sanger 1977

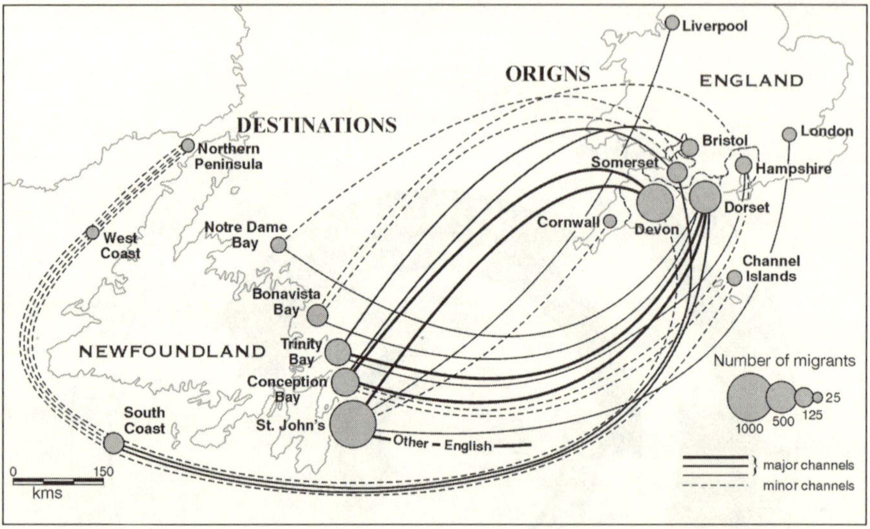

Fig. 1.3 Channels of emigration.
Source: Handcock 1989

Ill. 1.1 Flensing, traditional "pothead drive," Chapel Arm, Trinity Bay, Newfoundland, 1955 (R.E. Eshelman)

island as a local labour force for each fishing season.[21] By the early nineteenth century, resident merchants, mostly in St John's and Conception Bay, had taken over the market and supply roles previously controlled from England.[22] The resident population more than quadrupled from the 1780s to the 1820s and continued to grow rapidly.[23] When "modern" commercial whaling began in the late nineteenth century, virtually the whole population of Newfoundland and Labrador was native born. Over 200,000 people lived in more than a thousand small outport communities scattered along almost ten thousand kilometres of coastline.[24]

By this time almost 90 per cent of the expanding labour force was involved in salt cod production.[25] Unfortunately the industry could not keep pace with population growth. Exports of 29.2 quintals per capita in 1815 declined to just 5.6 by 1901.[26] Many fishing families became increasingly impoverished and turned to other subsistence activities.[27]

Some fishermen were able to drive ashore small whales such as potheads, *Globicephala melas* (ill. 1.1).[28] Many others also began to make annual trips "down north" to take part as "stationers" or "floaters" in the Labrador cod fishery, with some eventually remaining behind as "livyers."[29] A large number of Labrador fishers also began to use their schooners to search out harp-seal whelping patches on the offshore ice in early spring before they fitted out for the northern cod fishery.[30] Despite these adaptations, the traditional rural economy reached its limits of expansion by the 1880s.[31] Numerous attempts by the Newfoundland government to diversify and improve it were largely unsuccessful. The development of mines, a pulp and paper industry, and sawmilling had important effects on regional economies but was not able to create enough employment for the expanding population.[32]

Newfoundlanders began to emigrate to the "Boston States"[33] and "mainland" Canada. Many of those who remained endured hardship and deprivation which only began to be relieved in 1949 when Britain's oldest colony entered into confederation with Canada.[34] The introduction of the Canadian welfare system and government-sponsored resettlement and rural development initiatives greatly improved socio-economic conditions.[35]

From the mid-nineteenth century, however, these attempts to diversify and improve the rural economy largely failed to reduce dependency on the cod fishery. One initiative that enjoyed limited success in some localities was shore-based whaling, which used technology and expertise imported from Norway in the 1890s. Stations operated sporadically until the Government of Canada brought whaling to an end in 1972. These factories provided alternative employment and income for residents of nearby communities, resulting in commercial whaling entering the annual cycle of natural resources utilization in parts of rural Newfoundland and Labrador.

COMMERCIAL WHALING

Commercial shore-station whaling followed a pattern similar to that of most other whale fisheries (fig. 1.4).[36] The discovery of new hunting grounds led to a rapid increase in competition and catching effort. This quickly depleted whale stocks to unprofitable levels, and new whaling grounds had to be sought. Technological innovations enabled renewed hunting on previously abandoned grounds and initiated new whale fisheries based on previously unexploited species.[37] Continuing depletion of stocks and increasing condemnation of whaling resulted in the International Whaling Commission (IWC) placing a global moratorium on the industry in 1982.[38]

Traditional Hunting

Commercial whaling began in Newfoundland and Labrador in the sixteenth century when Basque fishers established stations in the Strait of Belle Isle as an extension of an industry that grew from activities in the Bay of Biscay.[39] The Basque are generally credited with elevating shore-based whaling above the small-scale subsistence level during the twelfth century (fig. 1.5).[40] This fishery was based exclusively on the North Atlantic right whale, *Eubalaena glacialis*, a timid and slow-moving animal that floated when killed, could be easily retrieved, and provided large quantities of meat, blubber, and baleen.[41] Harpoons were thrown from small, sail-assisted rowboats to impale the whales, which were then killed with lances and towed to shore for processing.[42]

The Bay of Biscay industry had declined significantly by the early sixteenth century, and the Basques almost certainly used information brought home by fishermen to transfer their whaling operation to eastern North America.[43] Seasonal stations were located at over a dozen sites along the Labrador coast (fig. 1.6). This fishery is well documented in Spanish archives[44] and confirmed by accounts from early English visitors. Anthonie Parkhurst, a native of Bristol who had made four voyages to Newfoundland, wrote on 13 November 1578 to the chronicler Richard Hakluyt, "I am informed that there are above 100. saile of Spaniards that come to take Cod … besides 20. or 30. more that come from Baskaie to kill Whale for Traine [oil]."[45] Similarly, Richard Whitbourne, a veteran of forty years experience in the migratory fishery, recorded on his first voyage to Newfoundland in about 1589 that "wee were bound to the Grand Bay (which lieth on the North-side of the Land) purposing there to trade then with the Savage people … and to kill Whales, and to make Traine Oyle, as the Biscaines doe There yeerely in great abundance."[46]

Basque whaling began in Labrador in the second quarter of the sixteenth century and expanded significantly in the middle decades. By the 1580s[47] it was in decline as a result of overexploitation, the development of more lucrative whaling at Spitsbergen, growing competition from Dutch and English whalers, domestic strife in Spain, and European wars.[48] However, aboriginal whaling continued along the shores of New England, in the Gulf of St Lawrence, perhaps on the Labrador coast to Hudson Bay and Davis Strait. French

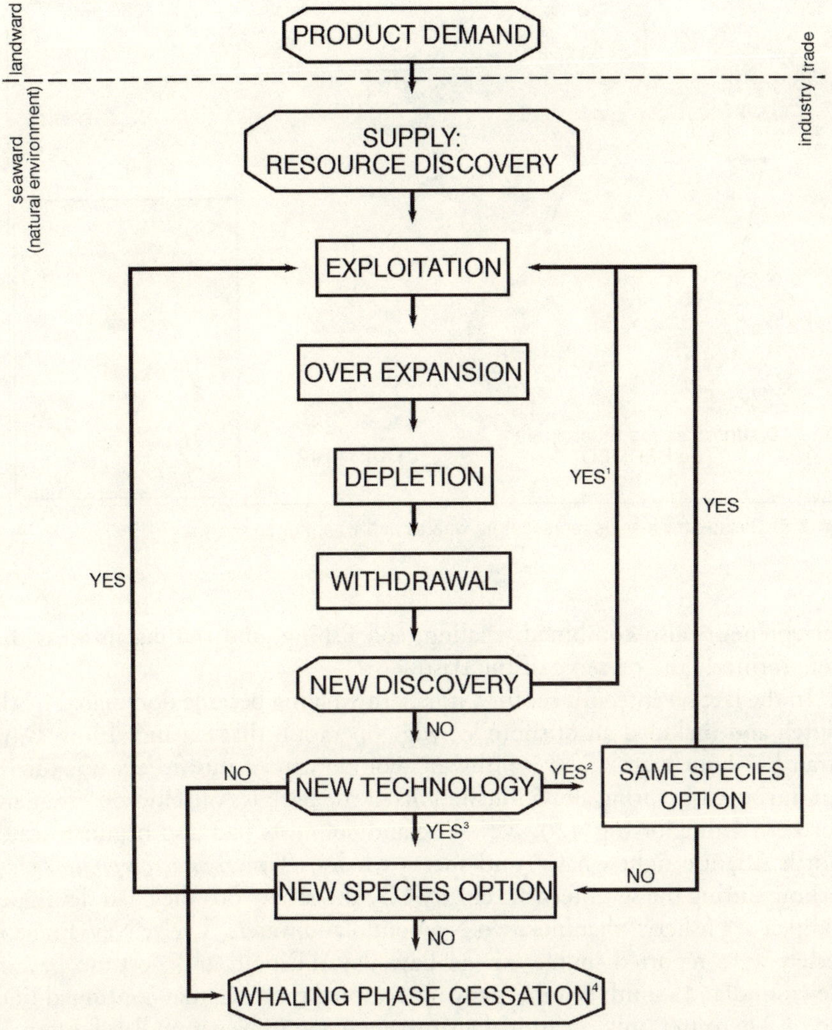

1. Spatial diffusion: same species or new species supplying same demands using existing technology.
2. More effective and efficient technology permitting continued exploitation of declining resource stocks.
3. Capture of species previously not available before adoption of new technology.
4. Appropriate to pre-twentieth century whaling. Although model does not incorporate resource management and conservation regulations which have influenced post World War I practices, similar patterns have characterized modern whaling.

Fig. 1.4 Whaling: pre-twentieth century historical patterns.

Source: Sanger 1985

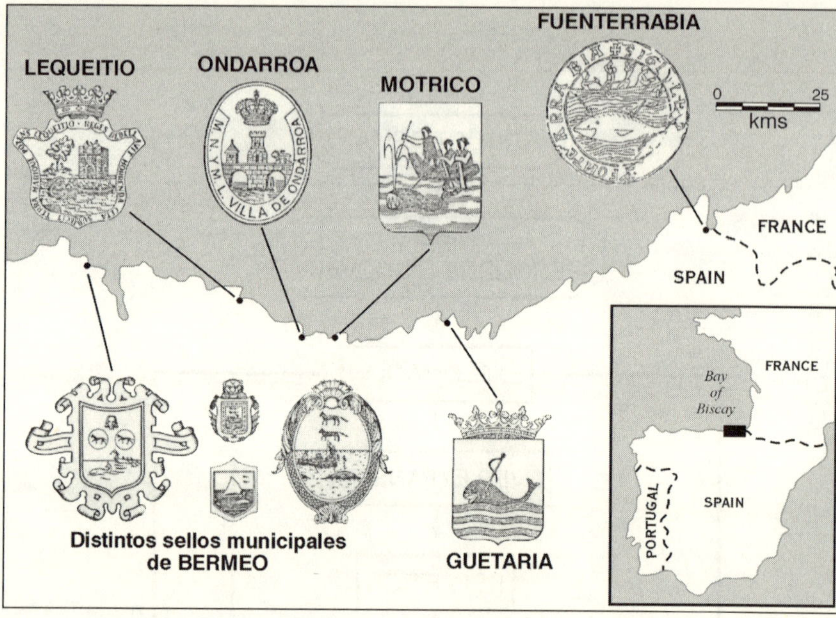

Fig. 1.5 Basque town seals representing whales and whaling. *Source*: Sanger 1985

entrepreneurs also combined whaling, cod fishing, and trading in areas that were formerly the preserve of the Basques.[49]

In the late seventeenth century, northern whaling became dominated by the Dutch and included an offshore ice-edge operation that expanded into Davis Strait by about 1720.[50] The "southwest" component of this fishery was carried out during early spring along the margins of the pack ice off Hudson Strait and northern Labrador (fig. 1.7). New England colonists had also begun to catch North Atlantic right whales, and sperm whales, *Physeter macrocephalus*, close inshore during the seventeenth century.[51] By the mid-1700s they had developed an open-sea fishery[52] that included Newfoundland waters. Twelve New England vessels were reported in Hermitage Bay, Bay d'Espoir, and Fortune Bay on Newfoundland's south coast as part of a seasonal presence that continued until 1807.[53] Thereafter only occasional visits were made to Newfoundland, often on the way to more lucrative hunting grounds in Hudson Bay, Davis Strait, and Baffin Bay.[54]

The growing nineteenth century resident population of Newfoundland restricted its whaling to driving animals ashore or using carcasses found stranded or floating at sea. The oil, bone, and meat were welcome supplements to meagre subsistence diets and incomes. James and John Rourke of Mall Bay, for example, found a dead whale in 1897 which they sold to M. Tobin of nearby St Mary's, who had it rendered into oil;[55] and the schooner *United Brothers* arrived in St John's on 18 May 1895 with 450 cwt of blubber and three casks of oil from five whales taken by residents of Little Codroy River on the west coast of the island.[56] Similarly, Messrs Boyle, Lee, and Connors of Harbour Grace, fishing at Long Island Tickle, Labrador, in 1892, "secured a splendid whale, and from

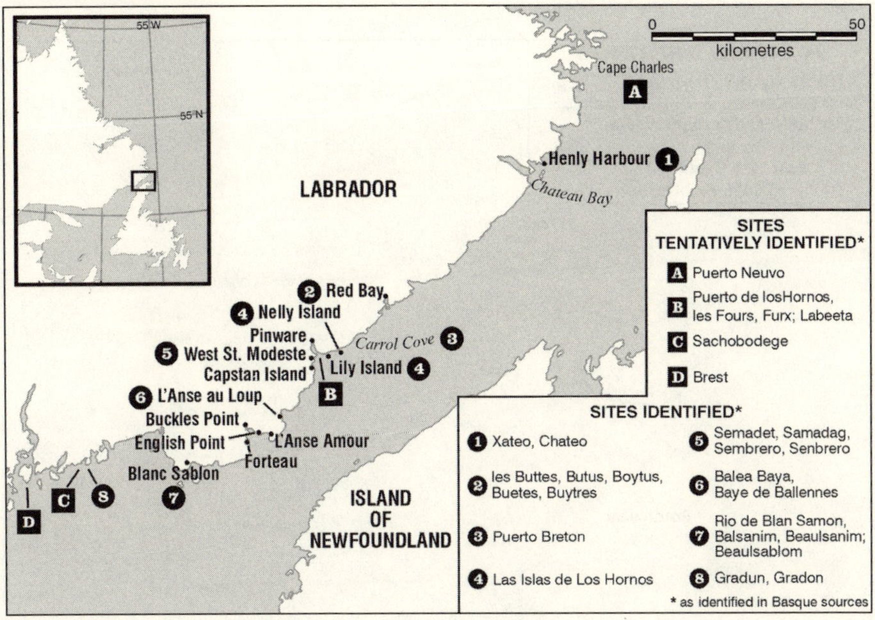

Fig. 1.6 Sixteenth century Basque whaling stations, Labrador.
Source: Sanger 1985, after S. Barkham

all that could be learned from residents of the locality, the bone and oil would net them a handsome sum."[57]

Several of the larger merchant firms involved in the salt-cod trade and sealing also conducted small-scale whaling during the nineteenth century, encouraged by legislation introduced in 1840 that offered a £200 bounty to each of the first three vessels landing at least ten tons of whale oil (or fifteen tons of blubber) before 10 November of that year. Two St John's companies, C.F. Bennett and Job Brothers, for example, employed whaling vessels along the south coast between 1841 and 1844.[58] The experience later enabled Job Bros. to play a significant role in the development of the modern local industry. A small whaling company was also established in 1840 at Gaultois in Hermitage Bay by Newman and Co. At its peak in the late 1870s, it captured forty to fifty whales annually.[59] The facilities and equipment were put up for sale in 1899, including "two whale boats and appurtenances."[60] Although some oil and bone was sold locally, most was exported. *Fanny* (Capt. Kidd), for example, left St John's for London in February 1895 with 788 casks of cod oil, 167 of seal oil, one of shark oil and six of whale oil.[61] Another voyage in May took twenty-three barrels of whale oil and 680 pounds of whalebone.[62]

This early local commercial whaling was facilitated by two Scottish companies, Alexander Stephen and Sons[63] and the Dundee Seal and Whale Fishing Company,[64] operating in St John's between 1876 and 1900. They hired Newfoundlanders to crew their vessels, which first went sealing before sailing

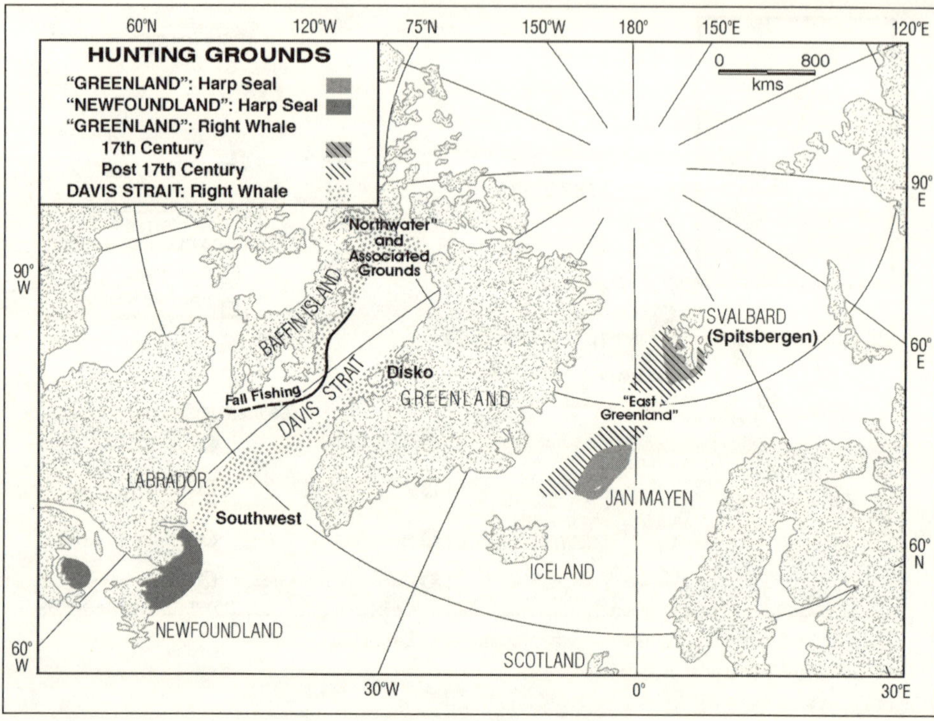

Fig. 1.7 Newfoundland and Scottish seal and right whale hunting grounds.
Source: Sanger 1985

north to hunt bowheads in Davis Strait and Baffin Bay.[65] The Scots also processed locally some of the whales they caught.[66] Both firms left Newfoundland just as investors were proposing to establish a domestic shore-based whaling industry. Machinery and experienced personnel became available to contribute to the construction and operation of new stations,[67] easing the transition of Newfoundland and Labrador small-scale traditional whaling into a "modern" industry.

The Modern Era

Right and sperm whale catches declined significantly by the second half of the nineteenth century, and the increased use of coal gas and petroleum products for lighting and lubrication reduced oil prices. More amenable and lucrative employment also became available in expanding onshore industries.[68] Nonetheless, there was enough demand for whale products to encourage brief revivals of some traditional whaling as new equipment and hunting practices were introduced. In the western Arctic, auxiliary steam-powered vessels (1880), shore-stations (1884), and over-wintering expeditions (1889) prolonged the traditional bowhead fishery.[69] These reduced severely depleted stocks even further. The industry was forced to turn its attention to the pursuit of unexploited species, ushering in the "Modern Era" of global whaling.[70]

Traditional whaling was based upon the use of open boats and non-explosive, hand-thrown harpoons. The faster-swimming and more powerful baleen whales (rorquals), which generally sank when dead, could not be caught. Although these were often abundant on traditional whaling grounds, experience showed that they "would drag down with them any boat whose steerer was rash enough to get fast; further, they were so active that they might tow a boat under or miles away from its ship, and they were difficult to kill."[71]

Attempts to replace traditional whaling became especially pronounced from the late 1850s[72] when an American, Thomas Roys, experimented with the use of rockets to kill whales.[73] The turning point in the development of the modern industry, however, was the perfection in Norway between 1864 and 1868 of a catching system that became the industry standard.[74] The *Spes et Fides,* the world's first purpose-built steam-powered whale catcher, was designed by sealing captain Svend Foyn of Tønsberg and launched in 1863. Foyn's whaling system eventually consisted of a single bow-mounted gun capable of firing an exploding harpoon attached to a winch and pulley mechanism strong and flexible enough to retrieve the largest whale (fig. 1.8 and ill. 1.2). Commercial whaling was radically transformed. The depletion of all commercially important whale stocks was now inevitable, especially since virtually all of the regulations introduced during the twentieth century, intended to prevent the modern industry from suffering the same fate as traditional whaling, were generally ineffective, often ignored, and poorly enforced.

Foyn introduced his catching system to the waters off Finnmark in northern Norway. By 1883 sixteen stations and twenty-three catchers were operating.[75] A rapid decline in coastal stocks and catches occurred, leading the Norwegian government to close the industry for ten years from 1 February 1904.[76] Norwegian whaling entrepreneurs were therefore forced to look elsewhere for unexploited stocks.[77] One such initiative led to the formation of a whaling company in Newfoundland and Labrador in 1896 as a partnership with local investors. Norwegians subsequently provided a significant amount of expertise, equipment, and financing for this new phase of Newfoundland and Labrador whaling.

The introduction of factory vessels in 1903 largely eliminated the need for shore stations and further hastened the global demise of the industry.[78] Whales could now be pursued throughout their migration routes and processed at sea. This method reached its zenith in the southern oceans after the first truly pelagic factory was introduced in 1925 but was rarely practised in the western North Atlantic. One Newfoundland whaling company, however, did send a fleet south in 1908–09. Other Newfoundland whalers gained southern oceans and pelagic experience after Britain's largest whaling company, Christian Salvesen, entered the local industry in the late 1930s and employed Newfoundlanders in its southern hemisphere operations.

Pelagic whaling was initially so successful that it produced an oversupply of whale oil, contributing to reduced prices and profits in the 1930s.[79] The industry again declined, including in Newfoundland and Labrador. A short-lived revival was interrupted by World War II, and strong postwar demand for

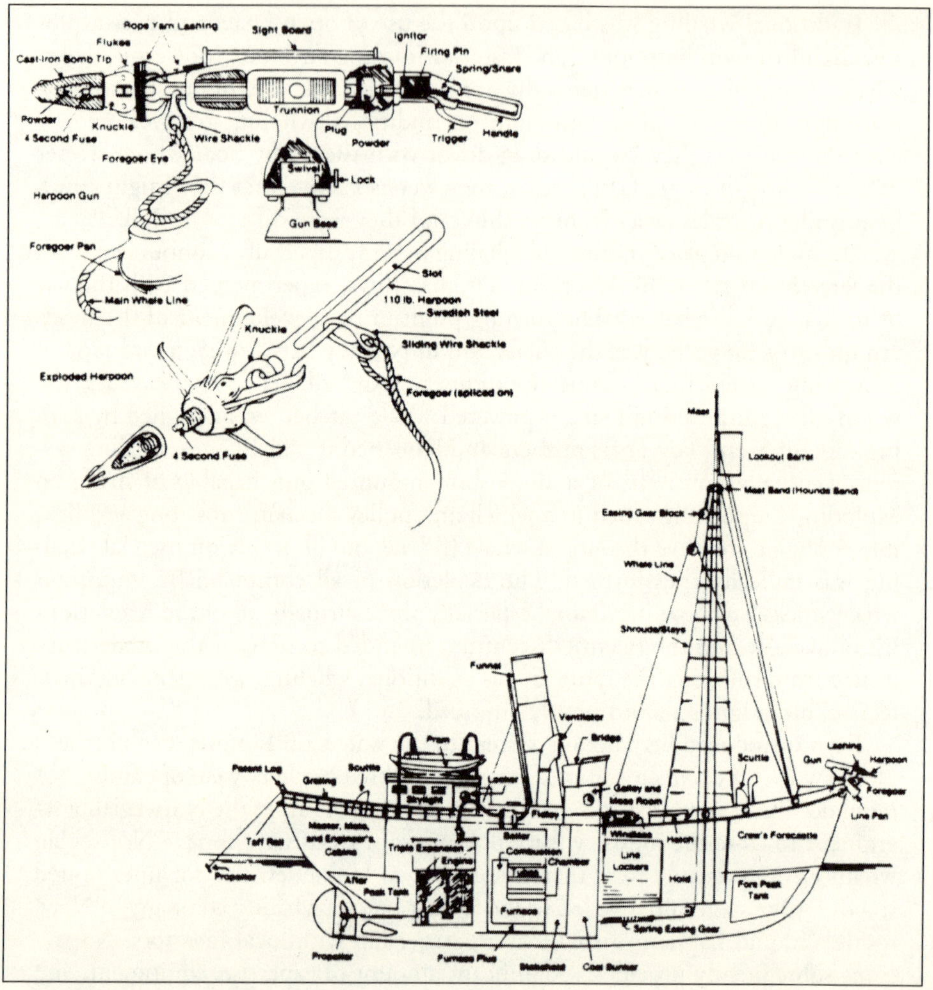

Fig. 1.8 Modern whaling technology *Source*: Hagelund 1987

edible fats again generated new interest.[80] Despite attempts at regulation by the newly formed International Whaling Commission (IWC), one species after another was overexploited.[81] The 1982 moratorium, implemented in 1986 for shore-stations, brought to an end most commercial whaling. Some continues under the guise of scientific whaling, or outside the jurisdiction of the IWC. The Government of Canada imposed its own moratorium in 1972, and the three stations operating, two in Newfoundland and one in Nova Scotia, closed.[82] Newfoundland and Labrador's sporadic involvement in commercial whaling had ended, more than four centuries after the Basques established their first seasonal stations.

Ill. 1.2 Modern technology, Japanese catchers, South Georgia, 1964 (A.B. Dickinson)

THE RESOURCE

Newfoundland and Labrador whaling companies were often poorly financed and equipped. Stations therefore had to be located close to stocks to reduce catching expenses and maximize profits. A knowledge of the migration habits that dictated the seasonal appearance of whales was critical. Although it is impossible to determine how many were taken by traditional whalers in the three centuries after the arrival of the Basques, records for the modern industry show that at least 19,643 were caught between 1898 and 1972 (appendix 1).

Species

The modern industry relied primarily on baleen whales (mysticeti). Fin whales, *Balaenoptera physalus*, predominated (12,863; 66 per cent), followed by blue, *Balaenoptera musculus* (1,450; 7 per cent), humpback, *Megaptera novaeangliae* (1,402; 7 per cent), minke, *Balaenoptera acutorostrata* (920; 5 per cent), and sei, *Balaenoptera borealis* (246; 1 per cent). Toothed whales (odontoceti) were occasionally taken, especially the sperm, *Physeter macrocephalus* (418; 2 per cent). "Others" (2,333; 12 per cent) included the long-finned pilot or pothead, *Globicephala melas* (324), killer, *Orcinus orca* (6), and Northern bottlenose, *Hyperoodon ampullatus* (1). One rare mysticete, the North Atlantic right whale, *Eubalaena glacialis,* was also captured. Large numbers of pothead whales were driven ashore annually, and a Norwegian fishery took 824 bottlenose, 142 minke, and 17 pilot whales from the Labrador Sea between 1969 and 1972 (table 1.1).

Fin whales concentrate primarily between the shore and 1,000 fathoms off Newfoundland and Labrador during the spring and fall.[83] Blues primarily occur seasonally along the south and west coasts of the island,[84] and humpbacks most often frequent eastern Newfoundland and southern Labrador coastal waters between early spring and late fall.[85]

Sei whales are oceanic and rarely move inshore. There may be two stocks in the western North Atlantic, on the Scotian Shelf and in the Labrador Sea. They usually arrive by early June, but their appearance is inconsistent;[86] few were thus processed. Sperm whales frequent the edges of the continental slope and deep ocean trenches, and those killed locally produced less than 1 per cent of the world's sperm oil.[87] Long-finned pilot (pothead) whales, one of the most common inshore species, appear in late spring and remain until the fall. They were increasingly hunted from stations after 1948 as catches of large whales declined. Most were processed at Dildo, Trinity Bay, to supply meat for a fur-farming industry. Large numbers were killed by local residents between 1948 and 1971 in onshore drives (table 1.2).

Minke whales, the smallest mysticete in Canadian waters, move inshore during the summer pursuing capelin, *Mallotus villosus*. Hunted by shore-station catchers after 1948 (appendix 1), their meat was also used by fur ranchers, or sold for human consumption as "arctic steak."

The seasonal presence of whales provided the opportunity for twenty-one whaling stations (appendix 2) to operate periodically between 1898 and 1972 (fig. 1.9). Although often short lived, they offered employment to local residents, allowing many to supplement meagre incomes gained from traditional sources. The industry was highly cyclical, largely as a result of over-exploitation and fluctuating market prices (fig. 1.10).

Table 1.1
Norwegian Catch of Small Whale Species off Labrador, 1969–72

Year	*Minke*	*Bottlenose*	*Pilot*
1969	9	237	–
1970	2	436	–
1971	11	151	–
1972	120	–	17
Total	142	824	17

Source: T. Christensen 1975.

PRODUCTS

Whales were hunted commercially primarily for their oil and baleen (whalebone).[88] The meat was also sometimes used for human consumption and animal feed, and other parts were converted into meal or "guano." The many items produced "added greatly for many centuries to the comfort and welfare of the civilized world."[89] Oil was the principal product initially, but whalebone became increasingly important towards the end of the eighteenth century and periodically throughout the nineteenth, when it "was nearly fourteen times more valuable than the oil, so that the Greenland whalers often took only the baleen and threw the rest of the carcass away."[90]

In the early days of the industry, oil was mainly used for making soft soap and for lighting,[91] but demand declined rapidly after coal gas and mineral oil products were introduced in the nineteenth century.[92] Markets grew again when it was discovered in the 1920s that whale oil could be solidified by hydrogenation and used to make high-grade soap and margarine.[93]

The method of extracting oil was essentially the same on all shore-stations. Carcasses were hauled onto a ramp (plan) where the skin and blubber were removed (flensed) using long-handled knives and winches. These were then cut into smaller pieces and boiled in cookers (digestors). In the modern industry, oil was initially extracted by steam only, but the introduction of pressure cookers improved yields (ill. 1.3). At the end of the cooking process, a mixture of oil, glue-water (an aqueous residue), and grax (solid particles) remained. The oil was cooled and stored, and the glue-water and grax were discarded. In later years these were treated in centrifugal separators to extract the last residues of oil. Anything left went into the production of bone and meat meal, or guano (ills. 1.4, 1.5).[94]

Table 1.2
Pothead Whale Catch, Newfoundland and Labrador, 1948–71

Year	*Catch*	*Year*	*Catch*	*Year*	*Catch*
1948	215	1956	9794	1964	2849
1949	0	1957	7831	1965	1520
1950	172	1958	789	1966	887
1951	3102	1959	1725	1967	739
1952	3155	1960	1957	1968	311
1953	3584	1961	6262	1969	123
1954	2298	1962	150	1970	155
1955	6612	1963	221	1971	4

Sources: Mercer 1975; Nelson and Lien 1996; Sergeant 1953 and 1962; Sergeant and Fisher 1957; Templeman 1966.

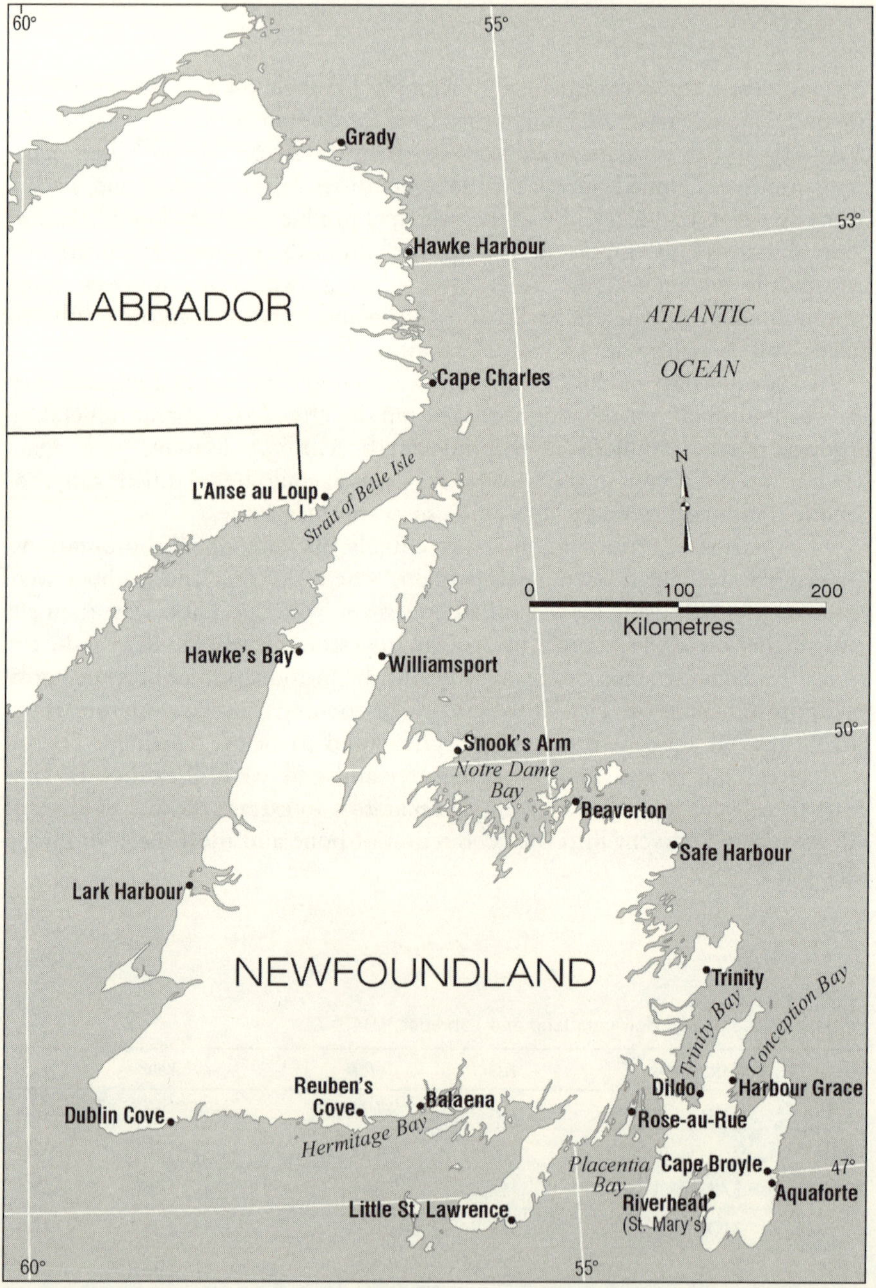

Fig. 1.9 Newfoundland and Labrador whaling stations, 1898–1972

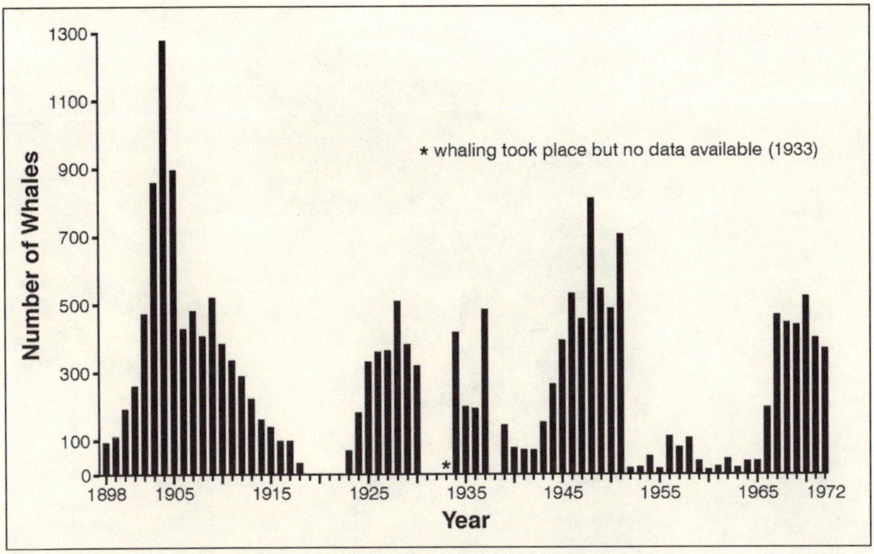

Fig. 1.10 Newfoundland and Labrador whaling catch, 1898–1972.
Source: IWC, Newfoundland, *Annual [Fisheries]* Reports, and newspapers (various).

The fortunes of a station depended to a large extent on its capacity to produce and sell high-quality oil in times of increased market demand. Quality was determined by free fatty-acid content, which in turn depended primarily on the degree of decomposition of the carcass. Grade 1 oil came mainly from the blubber and grades 2, 3, and 4 from meat, bone, and internal organs.[95] The objective of each station was to retrieve carcasses as fast as possible and produce high-quality grade 1 oil. Information is unavailable on the amount of the various grades of oil produced by the stations in Newfoundland and Labrador, but, at least in the early years, most was probably grade 1 since only the blubber was processed from whales killed relatively close to the factories.

For a time the baleen of the larger mysticetes was also an important product. Combining strength with elasticity, it was used by the fashion industry for such items as skirt hoops, stays, and hats. As women's fashions changed, however, and steel and plastics became more common, its economic value gradually diminished.[96]

In the early years of the modern Newfoundland and Labrador industry, flensed carcasses (skrotts) were cast adrift, often to float ashore again, to the annoyance of people living nearby. This, and the recognition that a significant portion of the resource and revenue was being lost, led to the introduction of legislation in 1902 requiring carcasses to be converted into guano instead of being discarded. The guano was often used as fertilizer or incorporated into animal feed. The revenue generated from its sale helped extend the operational lives of shore-stations.

Ill. 1.3 Rusting machinery, Aquaforte (A.B. Dickinson)

Ill. 1.4 Guano drier remnant, Hawke's Bay (A.B. Dickinson)

The cyclical nature and overall decline of the Newfoundland and Labrador modern whaling industry were caused primarily by a combination of stock depletion and price fluctuations. The first phase of modern times lasted just eighteen years and was followed by three clearly defined periods of revival and decline (fig. 1.10). Although it was legislation that ended the industry in 1972, it would probably not have lasted much longer because of declining profits and global concerns on the ethics of whaling.

Ill. 1.5 Guano production, Hawke Harbour, c. 1955 (J.C. Crosbie)

CHAPTER 2

Development

St John's newspapers carried detailed accounts of the early successes of the whaling captain Svend Foyn and his compatriots. The *Evening Herald*, for example, reported that in 1896 the Norwegian fleet had captured 792 "large" whales that yielded 3,200 tons of oil, selling for about £16–17 per ton in Glasgow.[1] Although some thought that a similar local industry would also generate substantial profits from a virtually unexploited resource, none of the Newfoundland companies involved in sealing or the Scottish Arctic whale fishery had the knowledge and skilled personnel needed to catch whales with the new methods. Introduction of the industry to Newfoundland and Labrador would require the development of joint ventures with experienced Norwegians.

The decline of Finnmark whaling had already encouraged Norwegians to seek opportunities elsewhere. They had expanded their activities westward by building stations in Iceland (1883) and the Faeroe Islands (1894),[2] a movement that reached Newfoundland in 1896 with the establishment of the Cabot Steam Whaling Co. Ltd. The company built a station at Snook's Arm (fig. 1.9) and began hunting in 1898. Not only was this an important step in the global expansion of the Norwegian industry but it also heralded the beginning of a new whaling phase in North America. Significantly, it coincided with a serious decline in the Newfoundland fisheries and a resulting economic depression.[3] Shore-station whaling was therefore able to provide alternative seasonal employment for some rural residents.

THE NORWEGIAN CONNECTION

The development of the new industry was facilitated by the presence in St John's of Adolph Nielsen, a Norwegian citizen with extensive fisheries experience. He was also familiar with the whaling techniques and operations of Svend Foyn. Nielsen had been involved in Norwegian attempts to raise cod larvae in hatcheries to restock their coastal waters,[4] and was one of two fishery inspectors sent to the United States and Canada in 1887 by the Norwegian government to study regional fishing industries.[5] During this trip he visited St John's and several fishing communities on the east coast of Newfoundland, and met Moses Harvey,[6] secretary of a commission of inquiry set up by the Newfoundland government to examine and report on the declining domestic fisheries.[7] Nielsen remained in contact with the commission, sending members a report describing the Norwegian fishery and its cod hatcheries.[8]

Returning to Newfoundland in February 1889, possibly at the invitation of the commission, Nielsen was subsequently appointed superintendent of the Newfoundland Fisheries Commission, a position established on 1 June 1889.[9] He became responsible for managing and revitalizing the industry. His efforts, including the construction of a cod hatchery on Dildo Island,[10] were rewarded when the government then made him the first superintendent of the Newfoundland Department of Fisheries, founded in 1893 from the Fisheries Commission.[11] A person with extensive fisheries knowledge and strong Norwegian connections was now strategically placed in Newfoundland to assist in the transfer of Norwegian whaling technology and expertise.

THE CABOT STEAM WHALING COMPANY

The first of these joint Newfoundland-Norwegian whaling ventures was established on 29 December 1896 as the Cabot Steam Whaling Co. Ltd. Its board of directors consisted of Adolph Nielsen and seven prominent St John's residents, including members of the Harvey family.[12] Nielsen's importance was reflected in his position as fourth largest shareholder (table 2.1).[13]

Nielsen's Norwegian background and contacts were soon put to good use by the board when they appointed him whaling manager.[14] Although in poor health,[15] he was sent to Norway to find financial, material, and technical support for the new venture. He was welcomed warmly by "the Norwegian capitalists [who] take a great interest in the enterprise, and will support any project recommended by Mr. Nielsen."[16] Investment in the company was clearly viewed as a profitable alternative to further involvement in the declining domestic whaling industry. Nineteen Norwegians held shares, compared to sixteen Newfoundlanders and one American (table 2.1), although Newfoundland residents, which included Nielsen, did in fact hold a small majority of the 551 foundation shares (56.1 per cent to 50.3 per cent). Norwegian investment

Table 2.1
Shareholders, Cabot Steam Whaling Co. Ltd, 19 January 1900

Shareholder	*Shares*	*Residence*	*Occupation*
Harvey and Co.	90	St John's	merchants
G.C. Hansen	51	Christiania	ship owner
K.R. Prowse	50	St John's	merchant
Adolph Nielsen	45	St John's	gentleman
Nielsen Bros.	30	Christiania	merchants
Hjalmar Bull	23	Christiania	captain
F.G. Hansen	23	Christiania	ship owner
William V. Whiteway	23	St John's	lawyer
H. Conrad	20	Christiania	ship owner
Halfdan Bull	18	Christiania	merchant
Ludvig Christiansen	18	Tønsberg	merchant
A.W. Harvey	16	St John's	merchant
Alex Harvey	15	St John's	merchant
Rolf Hansen	12	Christiania	ship owner
E. Jacobsen	10	Christiania	merchant
Hans Jacobsen	10	Frederickstad	doctor
Tønsberg Private Bank	10	Tønsberg	bankers
Aage Wettesen	10	Christiania	captain
John Harvey	9	St John's	merchant
Rev. Moses Harvey	9	St John's	clergyman
C.B. Isaaksen	9	Tønsberg	ship owner
H.R. Lange	9	Christiania	captain
M. Berbon	5	Tønsberg	agent
Mrs Emily Nielsen	5	Tønsberg	widow
Henry Bartlett	4	St John's	stevedore
E.T. Harvey	4	St John's	merchant
Charles Trippe	4	New York	telegraphist
Henry Blatch	3	St John's	wharfinger
Sigurd Isaaksen	3	Tønsberg	ship owner
H.M. Klocker	3	Christiania	agent
Capt. Bull	2	St John's	captain
James Norris	2	Green Bay	planter
P.T. McGrath	2	St John's	editor
R.G. Rendell	2	St John's	merchant
H. Feckudie [?]	1	Christiania	broker
Luke Pittman	1	New Perlican	schoolmaster

Source: Cabot Steam Whaling Co. Ltd, 1896–1914

Table 2.2
Share Distribution by Country of Residence, Cabot Steam Whaling Co. Ltd, 1900–14

	Jan. 1900	Sept. 1900	Dec. 1900	Dec. 1902	Dec. 1903	Feb. 1907	Feb. 1908	Feb. 1909	Feb. 1910	Feb. 1911	Feb. 1912	Feb. 1913	Feb. 1914
Newfoundland Resident Shares	277	366	366	366	310	310	300	325	343	343	343	339	340
Norway Resident Shares	270	260	282	282	342	342	352	385	367	367	367	371	370
United States Resident Shares	4	4	4	4	–	–	–	–	–	–	–	–	–
Total shares	551	630	652	652	652	652	652	710	710	710	710	710	710
Newfoundland Resident Shares as % of Total	50.3	58.1	56.1	56.1	47.5	47.5	46.0	45.7	48.3	48.3	48.3	47.8	47.9

Source: Cabot Steam Whaling Co. Ltd, 1896–1914.

increased slightly[17] (table 2.2) after Nielsen died in St John's of an asthma attack on 1 November 1903 and his sixty shares were transferred to his widow, Emily, in Norway.[18] Several Norwegians were subsequently elected to the board of directors, but Newfoundlanders continued to retain control.

An early priority of the company was to find an easily accessible site for their station, close to whale migration routes and a good supply of fresh water. Flensers and processing workers had to be hired and trained, catchers and crews obtained, and markets developed.[19] Nielsen was responsible for implementing all these local and overseas arrangements,[20] including the construction of a "completely equipped steamer [which] will be built in Norway this winter where alone they can be built."[21]

A suitable site for the station was found at Snook's Arm in Notre Dame Bay[22] (fig. 1.9), and the first oil production machinery and storage tanks were bought from A. Stephen and Son in St John's.[23] This Dundee shipbuilding company had extensive experience in the moribund Arctic whaling industry and in Newfoundland sealing but was reducing its local operations after a twenty-year involvement. As one St John's newspaper commented, "almost all the machinery required is here, at the dock, so that this will mean a large saving in the matter of purchase for the factory."[24] The closure of the Dundee Arctic Fisheries,[25] the only other Scottish whaling company involved in the Newfoundland seal fishery, provided the opportunity to purchase their hulk *Skimmer of the Waves*[26] to carry equipment, men, and stores to the construction site.[27] Additional machinery was bought for $12,500 from the insolvent local merchant firm of Thorburn and Tessier.[28] Preparations for the start of the new industry, therefore, were well advanced by the spring of 1898, and there were high expectations[29] of much-needed employment, especially since large numbers of whales had been reported to the south of St John's.[30]

WHALING OFF ST JOHN'S

The Cabot Steam Whaling Company had hoped to start construction at Snook's Arm in May 1898, but heavy ice along the northeast coast caused a delay.[31] *Skimmer of the Waves* did not leave St John's until 10 June with machinery and twenty employees from Angel and Company to assemble it.[32] Meanwhile, the newly built catcher *Cabot*[33] (table 2.3) left Christiania (Oslo) on 28 May 1898 under the command of Captain Henry Bull, an experienced Norwegian whaler and minor shareholder in the company. Sailing via Galway, Bull brought *Cabot* into St John's on 17 June, where it was declared "a handsome little vessel, fitted in every way for the prosecution of the whale fishery, and has triple expansion engines, her average speed being 12 knots. The cabin and forecastle are commodious and bright."[34] The ship was registered in St John's on 25 August 1898,[35] and the quality of the construction was noted by the local surveyors who transferred *Cabot* to the British flag. They reported that the ship "reflects the greatest credit upon the builder, and she is the very best they have ever put out of their hands."[36]

Although *Cabot* was supposed to sail north to Snook's Arm soon after arriving from Norway, the delay in building the station gave the company a chance to hunt the large number of whales reported off the Avalon Peninsula.[37] Captain Bull took *Cabot* into Freshwater Bay outside St John's harbour on 23 June, successfully tested the harpoon gun, and began whaling the following day.[38] The trip was intended primarily to familiarize the Norwegian crew with local

Table 2.3
Specifications of *Cabot*

Official No.	*108934*	
Registration		St John's, 1898
Built		Christiania, Norway, 1898
Construction		Wood/Steel
Dimensions	Length	86'1"
	Breadth	17'6"
	Depth (hold)	10'5"
Tonnage	Gross 98	Net 39
Horsepower		35
Registration closed	1951: Condemned, Harbour Grace, Newfoundland	

Source: *Mercantile List and Maritime Directory*, 1904.

conditions, but a fin whale was shot off Baccalieu Island on 26 June,[39] thereby initiating "modern" shore-station whaling in North America.

The carcass was brought into St John's the next day to be flensed at the premises of A. Stephen and Son, also bought by the company. Adolph Nielsen personally supervised the process. "The jaws of this monster of the deep are held open by tackles, and the inside is cut out, which contains the bone. Two planks are stretched across the lower jaw, and two men work comfortably in the opening. The fat is clear and about 10 inches thick in places, while the flesh is a bright salmon color and very attractive."[40]

As whale-oil production was the sole objective of the company, the flensed carcass was towed out to sea and set adrift. Unfortunately, it floated ashore at nearby Torbay, to the dismay of local residents who found "the aroma from it is not at all pleasant, and [it] suffuses the place."[41] Nevertheless, the fact that *Cabot* had captured a large whale, and sighted five more on the way back to St John's, prompted further optimism for the new industry which, it was hoped, would provide "employment to several hundred men for at least 9 months out of the year, if not longer."[42]

Although the carcass was flensed in St John's, oil could not be produced since the necessary equipment had been shipped to Snook's Arm. The company therefore stored the blubber on the chartered schooner *San Juan,* intending to transport it later to Snook's Arm for processing.[43] Meanwhile Nielsen was sent to Snook's Arm to speed up construction of the principal buildings, slipway, and twenty-three oil production and storage tanks.[44] Eleven of these were housed in a glass-roofed shed to increase the extraction temperature and accelerate the process.

Even before the station was completed, it became an important source of local income. Fifty men, mainly from nearby Tilt Cove, were hired to help the workers brought out from St John's. The government also added Snook's Arm to the schedule of their coastal boat *Alert,* contributing "significantly to the success of the enterprise" by guaranteeing regular access for personnel, supplies, and products.[45]

Captain Bull, meanwhile, made a second voyage, this time hunting at the entrance to Trinity Bay. Before *Cabot* had to return to St John's for coal on 7 July, seven more whales were taken. They were towed into New Perlican, a nearby outport community, where the crew of *San Juan* removed and stowed the blubber.[46] Bull was anxious to kill as many whales as possible and sailed again that same evening. This third voyage ended nine days later when *Cabot* returned to St John's, towing *San Juan* with the blubber of four more whales on board.[47] As the *Evening Herald* reported, *Cabot* had "done good work since leaving here, and in 16 days secured 11 whales, which with the one obtained before leaving, gives her a dozen to start with. Of this number 3 are 'humpbacked' a better and more valuable fish in many respects."[48]

Unfortunately, the stored blubber deteriorated and probably reduced the quantity and quality of the oil yield. This highlighted the importance of opening the Snook's Arm factory as quickly as possible.[49] Concern was also being expressed regarding the proper disposal of carcasses. According to one commentator: "The only thing those interested in the venture have to fear is an uprising of the people of the settlements where the drifting carcasses go ashore. Torbay, Scilly Cove and Hant's Harbour have each received such a visitation, and the inhabitants say the stench arising from the putrifying bodies is intolerable."[50]

The enthusiasm over the initial success of *Cabot* also overshadowed any concerns about overexploitation. Large numbers of whales continued to be sighted between St John's and Cape Race (fig. 1.1). The brigantine *Beagle*, for example, reported that "hundreds" were seen during passage from Harbour Breton.[51] Consequently the company decided to hunt from Aquaforte to the south of St John's.[52] There whales were so abundant that the experienced Norwegians killed them faster than they could be flensed by the *San Juan* crew. The company therefore chartered another vessel, *Helena May,* "so that the work of capture by the steamer may go on un-interruptedly." This unexpected investment was felt to be warranted since there was "every prospect of the company making a very good dividend on their outlay."[53] *Helena May*, however, was a poor choice since it was unable to hoist large pieces of blubber on board.[54]

Cabot took seven whales in two days off Aquaforte, and could have captured even more if Bull had not had to stop hunting so that his crew could help with flensing.[55] The rate at which whales were being taken elicited in the St John's *Evening Herald* the first suggestion that regulations would be needed to ensure the future of the new venture: "We may soon expect the inevitable 'crank' advocating legislation to prevent the killing of these huge monsters on the ground that the present industry will exterminate them."[56]

The success of the company was also noticed outside Newfoundland. The *Halifax Chronicle,* for example, called for local entrepreneurs to show similar initiatives: "The recent establishment of a whale oil manufactory in Notre Dame Bay, calls attention to the fact that whaling is now a paying venture off the coast of Newfoundland. Recently seven whales worth nearly $1,000 each were captured by a company of Norwegian whalers. There is no apparent reason why Halifax should not become an outfitting port for the whaling industry."[57]

COMPLETION OF THE SNOOK'S ARM STATION

Although whales were also reported as being numerous along the northeast coast of the island in July,[58] the company did not benefit since the Snook's Arm station was still not operational. They continued hunting from Aquaforte, and Captain Bull killed twenty-six whales by 18 August, when they became scarce.[59] Finally the station was completed, and *Cabot* sailed north with *San Juan* and *Helena May* carrying blubber to be processed.[60] Bull continued to be successful, killing seven whales during three days' hunting in Trinity Bay.[61] The company was now in a position to function as originally intended, and,

Ill. 2.1 *Cabot* at Snook's Arm, whale in tow, c. 1900 (Smithsonian Institution)

Ill. 2.2 Humpback on slipway, Snook's Arm, c. 1900 (Smithsonian Institution)

encouraged by its success, the directors made plans to expand. The *Evening Herald* reported,

[They] are determined to do as much work as possible in the capture of whales next season and we understand Capt. Bull of the S.S. *Cabot*, shortly goes to Norway for the purpose of purchasing two if not three steamers similar to the one he commands, for use on this coast. The Snook's Arm factory is now almost completed, and will be in operation next week. The present system of securing the fat is not very profitable as the schooners carry no tanks, and a large quantity of oil has therefore been lost.[62]

Ill. 2.3 Flensing humpback, Snook's Arm, c. 1900 (Smithsonian Institution)

Ill. 2.4 Cutting blubber for boiling, Snook's Arm, c. 1900 (Smithsonian Institution)

The Snook's Arm station became fully operational in the last week of August 1898[63] (ills. 2.1–2.4), and people came "in scores … to gaze in wonder at these monsters of the deep, the existence of which they never knew."[64] This success continued well into September.[65] In the station's first month of operation, *Cabot* had brought over twenty whales into Snook's Arm, "fin-backs and hump-backs, yielding from four to five tons of oil per whale," according to the *Evening Herald*. "Later on it is expected that the sulphur-bottom [blue] whales will make their appearance. These yield about twenty tons each. The manufacture of the oil at Snook's Arm goes on successfully under the skillful superintendence of Mr. Nielsen. They are now able to manufacture the fat of two whales daily. The prospect of taking whales continues excellent."[66]

Deteriorating weather conditions and the growing scarcity of whales as they moved south to their winter feeding grounds[67] led to the closing of the station in mid-November.[68] *Cabot* arrived back at St John's on 28 November with *Skimmer of the Waves*, marking the end of the first season. Ninety-one whales had been processed.[69] The local press praised the new technology and the joint venture and expressed further optimism for continued growth. Proclaimed the *Evening Herald*, "it will be no strange item in our exports in future, for the manufacture of this oil gives employment to a number of men; all hope the venture will meet with success."[70]

However, despite good catches and increased employment, whaling did not receive an unqualified endorsement from many of those residing near Snook's Arm. Concerns now began to be expressed that the traditional fisheries would be disrupted. It was reported that "the people of Green Bay will petition the government to prohibit the killing of whales in its waters, on the ground that the destruction of them will prevent herring and caplin from striking in to the shore, as the prevailing opinion among the fishermen is that the little fishes are driven to the shore by the whales."[71]

The first shipment of oil left St John's for Halifax in the *Damara* on 4 October 1898, totalling "520 casks, containing near 100 tuns of oil … the product of 40 whales [less than half of the season's total]."[72] In the absence of detailed production figures, income generated by the company during its first year is difficult to assess. Nevertheless, the Newfoundland Fisheries Commission listed 211 tons of oil produced, valued at $14,439, and whalebone at $1,167. This was a substantial increase from $1,581, the total value of the previous season's incidental whale catch (27 tons of oil at $1,340, bone at $241).[73]

Although the company had initially committed to having experienced Norwegian whalers train Newfoundlanders in hunting and catching techniques, this did not occur. Local workers continued to carry out the lower-paying menial and often repugnant processing tasks onshore. Working conditions left much to be desired, with the floors of the factory and "steep slanting wharf literally swimming in whale fat and the whole building dripping with whale oil. The men work in oil skin suits with rubber hats, and their faces shine and clothes ooze out the whale oil."[74]

The company also briefly tried to increase profits by producing whale meat for human consumption, since the men employed at Snook's Arm were already "using it freely and find it wholesome and nourishing … numerous vessels are coming in and carrying away huge chunks of it to be salted for winter use."[75] But this initiative was short lived, probably because revenue from oil sales made it imprudent to transfer labour to less profitable tasks.

The new industry did attract entrepreneurs who felt that they could make money processing the meat and by-products. In October 1898, for example, Messrs C. Doull and J. Shaddick arrived in St John's from Liverpool. Shaddick was the son of the general manager of the newly established Newfoundland Fish, Oil and Guano Company, and Doull was their Newfoundland agent.[76] They proposed building factories at Snook's Arm and elsewhere to produce frozen bait fish and squid, fish oil, and whale guano. Carcasses would be bought from the whaling company, minced, and the remaining oil extracted, filtered, and evaporated in a vacuum to the consistency of molasses. Salt and sugar were to be added for preservation and flavour, and the mixture bottled. The residue was to be ground and dried to become meal (guano). Both extract and guano were intended primarily for markets in the United States.[77] This proposal was not implemented, however, and the company went bankrupt in the summer of 1901.[78] Nevertheless, whole-carcass utilization and guano production were eventually required by the government.

Some also believed that whale oil had medicinal powers and that Snook's Arm could become a sanitarium:

> It is well known that where whale oil is manufactured no infectious diseases will come; and people with chest troubles are benefitted by inhaling the fumes and often cured of pulmonary affections … At that part of Dundee where whale oil is manufactured epidemic diseases are unknown. It was free from cholera when all around were dying; and invalids come to inhale the fume laden atmosphere. It would be curious if Snook's [Arm] should become a health resort, with whale flesh as the staple diet.[79]

Unsurprisingly, this did not come to pass.

Catch sizes during the first season suggested that a profitable and expanded industry was possible. It was apparent to the company that variations in local environmental conditions and migration patterns allowed for different hunting seasons off the north and south coasts. The Cabot Steam Whaling Company thus decided to increase catches and profits by expanding operations to the south coast in 1899.

CHAPTER 3

Expansion and Consolidation 1898–1902

The apparent abundance of whales and the ease by which Captain Bull had killed them convinced the Cabot Steam Whaling Company to expand operations. The first season had shown that Snook's Arm was too far from the spring and early summer whale migration routes, including that of the high-yielding blue whales; also, the northern bays usually became ice-bound late in the year. A second station was therefore built on the island's south coast (fig. 1.1) to extend the hunting season.

INITIAL EXPANSION

Cabot sailed from St John's in early December 1898 so that Captain Bull (ill. 3.1) could assess the availability of whales and search for a suitable construction site.[1] Placentia Bay was initially considered,[2] but Bull reported that prospects looked more favourable further west, and the company chose instead to build the station near McCallum in Bonne Bay, an inlet at the mouth of Hermitage Bay (fig. 1.9). The site had relatively flat land, a reliable fresh-water supply, and a deep-water anchorage[3] and would provide easy access to migrating whales during winter and early spring. The new station, to be known as Balaena (ill. 3.2),[4] would enable the company to carry out year-round whaling and, it was projected, catch more than two hundred whales annually.[5] Construction, and the purchase of a second catcher, were to be financed from an additional share issue[6] in the hope that the operation "would equal Snook's Arm next season."[7]

The new venture began when *Cabot* left St John's on 30 December 1898 for Hermitage Bay,[8] towing *Skimmer of the Waves* with workers, equipment, and building materials on board.[9] *San Juan* also brought coal to the site from Cape Breton.[10] Construction was slowed by exceptionally severe winter weather,[11] and an additional six machinists and four boilermakers were despatched from St John's in June by Angel and Co. to speed up the work.[12] Eight more employees arrived in September and "fitted up all the machinery for the whaling factory having quite a job. Twelve oil tanks that can contain thousands of gallons were put up and this will mean a deal of employment."[13] Adolph Nielsen also arrived to oversee construction of this station and have it ready for the winter fishery.[14]

Cabot hunted near the mouth of Hermitage Bay during the construction period, and on 2 February 1899 took the first blue whale of the local modern industry. Conditions were far from ideal:

> A heavy gale was blowing; the fierce cold froze every wave as it broke over the vessel, loading her with ice. The windlass, cables, even the gun platform were frozen. In spite of all Capt. Bull aimed so truly that he killed the big sulphur-bottom ... at a single shot and secured it. He saw some 20 or 30 more whales of the same kind, and by this time has probably captured more. Wonderful whalers are these Norsemen! It is evident that this is the headquarters of the most valuable species of whale around our coasts.[15]

With a potential oil yield of $750, about "three times as much as a hump-back or fin-back," the capture of such a large whale caused excited comment in the press, and expectations continued to grow for the successful expansion of the industry.[16] Bull killed eleven blues in the next two months,[17] but because the Balaena station had not yet been completed, the carcasses again had to be towed to a convenient flensing site and the blubber stored until it could be processed (ill. 3.3).[18] Oil yields and quality were therefore probably inferior.

Although whales continued to be numerous on the south coast,[19] *Cabot* was forced to return to St John's for repairs at the end of April.[20] She did not sail again until late June, this time to begin her second northern season supplying Snook's Arm.[21] It was also intended that Bull would later resume operations from Balaena when Snook's Arm closed for the season.[22] Meanwhile, continued sightings of whales along the south coast throughout the summer convinced the company to build a second catcher in Norway to be used full time at Balaena in 1900.[23]

The company also revised its hunting strategy for Snook's Arm. Instead of killing whales at every opportunity, it was decided to catch just two a day so as not to exceed the processing capacity of the station (ills. 3.4, 3.5).[24] Bull could now select larger individuals. Forty-three were killed (ills. 3.6, 3.7, 3.8) by mid-August, yielding 150 tons of oil valued at $80 per ton.[25] In September, 228 barrels of oil (six barrels per ton) were sent to St John's on *Virginia Lake*, with a further 289 barrels and 191 bags of whalebone arriving in October.[26] Two hundred barrels of the oil were shipped to Glasgow, an increasingly important international oil market.[27]

The "booming" 1899 season ended on 21 October with the return of *Cabot* to St John's. Bull had taken 109 whales[28] (ninety-eight from Snook's Arm, eleven from Balaena) for 211 tons of oil.[29] Individual yields, however, were low – only 11.6 barrels per whale (table 3.1) compared to an average of 29 barrels per

Ill. 3.1
A.W. Harvey and Captain Bull on *Cabot*, Notre Dame Bay, 1899 (Smithsonian Institution)

Ill. 3.2 The Cabot Steam Whaling Company's second whaling station, Balaena, Hermitage Bay (PANL A17-45)

Ill. 3.3 Blue Whale on slipway, Balaena, 1903 (Smithsonian Institution)

whale in 1904 from the first sub-Antarctic land station on South Georgia.[30] The difference was due primarily to catch composition, processing inefficiencies, and oil loss from stored blubber. Nevertheless, production and catch values significantly improved over the first season, good news for the future of the industry. Information on company profits is not available, but with $55,100 in share capital,[31] operating costs for two seasons, and sizeable expenses incurred in building one catcher and two stations, there was probably little margin.[32] Local investors may therefore not have been totally convinced of the viability of the industry and bought fewer shares in the company than the more knowledgeable Norwegians.[33]

As in the previous season, those living near Snook's Arm complained about odours from decaying carcasses that drifted inshore after being towed out to sea and abandoned.[34] A St John's newspaper reported that the problem extended to fishing and shipping: "it has been noticed that the carcasses of whales, from which the blubber has been taken, are cast adrift and float around Notre Dame Bay, and a few days ago one drifted into Twillingate, damaging herring nets. These are very dangerous to shipping, and some other means of disposing of the carcasses should be found."[35]

The company's third season began when *Cabot* left St John's for Hermitage Bay on 1 November 1899.[36] Whales were initially scarce, and Bull was unable to kill any during his first two cruises.[37] Hunting improved as more migrated into the bay, and five were taken before the end of December.[38] Eighteen more were killed in the first three months of 1900,[39] and eighty-nine by mid-May, keeping the men at Balaena "busy night and day cutting them up."[40] Whales became

Ill. 3.4 Captain Henrik Ellefsen firing harpoon from *Humber* off Aquaforte, c. 1904 (Ellefsen Papers)

Ill. 3.5 *Cabot* towing fin whale to Snook's Arm, 1899 (Smithsonian Institution)

Ill. 3.6 Fin whale on slipway, Snook's Arm, 1899 (Smithsonian Institution)

Ill. 3.7 Partially flensed fin whale, oil barrels, and *Cabot*, Snook's Arm, 1899 (Smithsonian Institution)

Ill. 3.8 Blubber and whalebone (baleen) from fin whale, Snook's Arm, 1899 (Smithsonian Institution)

Table 3.1
Catch and Production, Newfoundland and Labrador Shore-Stations, 1898–1901

Season	*Catch*	*Oil (Tons)*	*Barrels per whale*	*Oil ($)*	*Whalebone ($)*	*Total ($)*
1898	91	27	1.8	1,340	241	1,581
1899	109	211	11.6	14,439	1,167	15,606
1899–1900	190	430	13.5	34,604	1,824	36,428
1900–01	258	633.5	14.7	54,221	13,550	67,771
Totals	648	1,301.5	12.0	104,604	16,782	121,386

Source: Newfoundland, *Annual [Fisheries] Report for 1903*, 1904.

less common thereafter, but Bull killed another fifteen before returning to St John's on 29 July, bringing the south coast hunt to a close. In total 104 whales, "mostly small fish," were captured.[41] The first half of the 1899–1900 season was deemed a success, and the company decided to improve Balaena's wharf and storage facilities.[42]

Cabot left St John's for Snook's Arm on 6 August to initiate the northern fishery, returning to at the end of October after taking another eighty-six whales.[43] The strategy of hunting different stocks at different times of the year from stations on the south and northeast coasts proved highly successful, yielding 190 whales.[44] Revenue from oil and bone sales more than doubled over the previous season (table 3.1) due to improved yields, values, and demand.[45] For the first time the company made a small profit.[46]

CONSOLIDATION: THE SECOND WHALING COMPANY

Although it had taken the Cabot Steam Whaling Company three seasons to establish a viable operation,[47] the prospects of earning huge profits encouraged other entrepreneurs to invest in the industry. A second company, the Newfoundland Steam Whaling Co. Ltd, was formed on 15 September 1900 with $44,500 capital (tables 3.2, 3.3). Shareholders initially decided to build their station next to Balaena but quickly realized that this would only increase local competition.[48] Another site was chosen further west at Reuben's Cove in Chaleur Bay (fig. 1.9), and construction began immediately so the station could be ready for the 1900–01 season.[49] The schooners *Lizzie Bennett* and *Evadne* sailed from St John's with building materials and coal,[50] and two boilers were shipped to the site in *Greenland* on 22 April 1901.[51] Other stores and equipment arrived in the schooner *Rejoinder* on 3 June 1901, including a five-thousand gallon oil storage tank.[52] The station was finished by the end of June, and arrangements were made to have oil production supervised by Dr Ludwig Rissmüller.[53] The German-American chemist and engineer, according to a writer of the times, had "done more for whaling and the use of whale products than any other living man. To him is owed the utilisation of every part of the whale, including the flesh, the blood and liver, and parts of the skin which were only regarded as wastage a few years ago."[54]

The company's first catcher, *Puma* (table 3.4), was built in Christiania (Oslo) under the supervision of company secretary Alexander McDougall,[55] and arrived in St John's on 15 June 1901. It sailed for Chaleur Bay the same day and began hunting on 20 June.[56] Two whales were killed on the first voyage, "not bad to start with and all hope she will do well."[57] *Puma* continued to operate into the fall, capturing forty-eight whales.[58]

Cabot, meanwhile, required an extensive dry-dock overhaul in November 1900 before sailing for Balaena for the southern phase of the Cabot Steam Whaling Company's fourth whaling season.[59] Early reports indicated that whales were

Table 3.2
Founding Shareholders of the Newfoundland Steam Whaling Co. Ltd, 15 September 1900

Shareholder	*Address*	*Occupation*	*No. Shares*
Robert K. Bishop	St John's	merchant	80
Alexander McDougall	St John's	merchant	60
John Browning	St John's	merchant	50
Alan C. Goodridge	St John's	accountant	50
Daniel Murray	Glasgow	merchant	50
Marmaduke G. Winter	St John's	merchant	50
Thomas B. Goodridge	St John's	accountant	25
Baird, Gordon and Co.	St John's	merchants	20
Alan Goodridge and Sons	St John's	merchants	20
(?) Giraud	St John's	merchant	10
John C. Hepburn	St John's	accountant	10
Francis J. Hunter	St John's	bank manager	10
Michael Thorburn	St John's	merchant	5
Mary Grant Urquhart	St John's	bank manager	5

Source: Newfoundland Steam Whaling Co. Ltd, 1900–10.

Table 3.3
Shares and Capital of the Newfoundland Steam Whaling Co. Ltd, 1900–10

Date	*No. Shares Held*	*Nominal Value ($)*
15 September 1900	445	44,500
5 February 1902	500	50,000
1 February 1903	750	75,000
4 February1904	750	75,000
23 March1904	Issue increased to 2,500 shares @ $100	
31 December 1904	1,220	122,000
2 February 1906	1,275	127,500
17 February 1909	2,470	247,000
10 February 1910	2,470	247,000
2 March 1910	Receivership	

Source: Newfoundland Steam Whaling Co. Ltd, 1900–10.

scarce,[60] but numbers increased in May, and Bull was able to capture seventy-four by 20 June.[61] More could have been killed, but the factory was only able to process two daily.[62] A month later whales became scarce "as the caplin are leaving the shore."[63] *Cabot* returned to St John's on 28 July having killed 103 whales along the south coast, one less than the previous season.[64] Bull then sailed to Snook's Arm on 5 August,[65] and by 1 October the station "had all the whales the men could handle."[66] By the time the 1900–01 season ended in November, *Cabot* had delivered 102 carcasses to Snook's Arm. During this fourth and best season so far, the Cabot Steam Whaling Company processed 210 whales.[67] This, together with the forty-eight killed by *Puma* for the Newfoundland Steam Whaling Company, increased the total catch[68] and industry value by 36 per cent and 86 per cent respectively over the previous season (table 3.1).

INTRODUCTION OF REGULATIONS

The rapidly growing industry generated products valued at $121,386 in four years (table 3.1) and now began to attract significant attention from local and foreign investors. The measure of this interest is shown by the results of a public auction of thirteen $100 shares in the Cabot Steam Whaling Company. It had just paid dividends ranging between 30 and 40 per cent, and "brisk" bidding caused individual shares to sell "for $230, being at a premium of 130 per cent."[69] Shore-station whaling was now clearly perceived as an attractive investment opportunity. Other companies were formed, and plans were announced in late 1901–early 1902 for new stations at Aquaforte and Cape Broyle on the Avalon Peninsula, and Rose-au-Rue in Placentia Bay (fig. 1.9).[70]

This growing interest, however, began to arouse concern among members of the public,[71] who feared that uncontrolled expansion would destroy stocks, as had occurred in Norway. Shareholders in the two established companies

Table 3.4
Specifications of *Puma*

Official No.	*113823*	
Registration	St John's, 1901	
Built	Christiania, Norway, 1901	
Construction	Steel/iron	
Dimensions	Length	88'
	Width	17'
	Depth (hold)	10'3"
Tonnage	Gross 98	net 94
Horsepower	37	
Registration closed	1916: Sold to foreigners	

Source: *Mercantile List and Maritime Directory*, 1901.

also worried that increased competition would reduce their profits, contending that there was nothing "to prevent Norwegians from coming here and destroying the industry, which they could easily do as they would have no permanent interest in the colony or in the judicious prosecution of the business."[72] They argued that local citizens should have both more control and increased benefits. As a result the government was asked to restrict foreign ownership to 50 per cent of the shares.[73] Premier Sir Robert Bond acknowledged that the government wished to "conserve the interests of the industry"[74] and announced in the Throne Speech of the 1902 Session of the House of Assembly that "the Whale Fishery, which for some years was entirely abandoned, is now being vigorously prosecuted by two local companies, and with marked success. It has appeared to my Ministers that some Legislation is necessary in order to govern the prosecution of this industry, and a measure will be introduced for that purpose."[75]

Bond was careful to point out that the proposed whaling legislation was "for the preservation of the whaling industry, not for the maintenance of any monopoly."[76] The bill received considerable debate following the Throne Speech.[77] Although the government continued to declare that it intended only to "regulate the conduct of the industry," several members argued that the interests of local investors should be promoted. In the end there was agreement that the industry would benefit from an infusion of local and foreign investment capital.[78]

The ordinary outport fishermen, meanwhile, had more basic concerns. They feared that whaling would harm the local cod, herring, and lobster fisheries. Many also felt that reduced cod catches in the vicinity of Balaena were due to fewer whales being available to drive fish inshore.[79] Residents of Red Island, Fox Harbour, and Cape La Hune petitioned the House to introduce legislation to reduce possible disturbance to marine life and the environment from decaying carcasses, floating oil, and station and catcher refuse.[80] Their objections were supported by the opposition, which demanded a commission of inquiry to examine the relationship between whaling and the inshore fishery.[81]

The government was sympathetic, but rejected the notion that whaling might adversely affect the inshore fishery and voted against a commission. This was perhaps to be expected, since several government members and their influential friends were shareholders in the two companies. They argued that whaling did not appear to have affected the inshore fishery of Norway, "the most advanced country in the world in the matter of the scientific prosecution of the fisheries."[82] Significantly, whaling was also now seen as a potentially important source of income for the Treasury.

The first reading of the "Act to Regulate the Whaling Industry" was given on 15 March 1902.[83] The bill established regulations for station construction, hunting, prevention of pollution, and protection of the inshore fishery.[84] Each new company had to advertise in the government's *Royal Gazette* and at least one local newspaper its intention to build a station, and was required to purchase a ten-year operating licence at a cost of $2,000 annually. The high dividends recently earned by shareholders in the Cabot Steam Whaling Company suggested that this was a fair assessment. Furthermore, licences would not be awarded until the government had approved each company's operating intentions and its station had been built. Factories were to be located at least fifty miles apart to reduce competition for local whale stocks and to minimize processing odour. Individual stations could only be supplied by one catcher, and towboats were prohibited to prevent whalers from remaining on the grounds for long periods and killing excessive numbers. Hunting was also banned within one-half mile of a sailing vessel to reduce possible interference with fishing, and companies were financially liable for any damage to vessels and equipment. Whaling without a licence was penalized by fines up to $1,000 and the possible confiscation of vessels and equipment.

In response to complaints about environmental pollution, each factory was required to convert offal and flensed carcasses into guano.[85] Concerns about Norwegian domination were counteracted by making it illegal for companies to employ onshore workers who were not British subjects or had not been Newfoundland residents for at least two years. Foreign crews, however, could continue to be employed on catchers for three years to train local seaman.

Considerable debate ensued in the legislature.[86] In response to charges that the act would place control of the industry in the hands of a few "capitalists" who could afford the licence, the government reduced the annual fee to $1,500 per station. They also agreed to an amendment whereby the Cabot Steam Whaling Company could deploy their two catchers from either Snook's Arm or Balaena, since both had been built prior to the introduction of the new legislation.

The amended act was passed on 22 April 1902[87] and became the first regulatory framework for the modern Newfoundland whaling industry. Its impact was also felt further afield, providing a model for a 1904 Amendment to the Canadian Fisheries Act which governed the growing British Columbia industry[88] and for the subsequent regulation of whale fisheries elsewhere in the British Empire. The introduction of licences and fees and the regulation of catcher numbers became the basis of the Whale Fishery Ordinance of 1906, the first legislation introduced by the British government to control the rapidly expanding industry in the Falkland Islands Dependencies.[89]

Although the 1902 act was intended to conserve stocks and sustain shore-station whaling in Newfoundland and Labrador, it had serious omissions that prevented these objectives from being attained. Killing of females was not prohibited, thus adversely affecting stock recruitment. In 1899, for example, 60 per cent of twenty-five fin whales sampled at Snook's Arm were female, as were 37 per cent of ninety-two blue whales examined at Balaena in 1900–01. Furthermore, 66 per cent of the female fin whales were pregnant.[90] The act also did not limit the number of licences that could be awarded or identify a maximum annual catch per licensee. Consequently, the incorporation of new companies and the building of additional stations continued at a rapid pace, with disastrous consequences for both the stocks and the industry.

CHAPTER 4

Expansion, 1902–03

THE 1902 SEASON

The industry continued to grow in 1902, stimulated by reports of large numbers of whales, particularly off the eastern Avalon Peninsula.[1] These waters had last been hunted in 1898 by the Cabot Steam Whaling Company as they waited for their station at Snook's Arm to be built. The company now decided to return to these grounds by entering into a partnership with Bowring Bros. of St John's as the Cape Broyle Whaling and Trading Co. Ltd.[2] Land was bought at Admiral's Cove, Cape Broyle (fig. 1.9), and construction of the new station began in March 1902 under the supervision of Michael P. Cashin, a local merchant and major shareholder.[3] Processing was scheduled to begin in July but was delayed by construction problems.[4] A forty-foot wharf was completed by the end of September, but the factory was not finished that year.[5]

Local investors continued to dominate the expanding industry, purchasing 79 per cent of the first share offer (table 4.1).[6] Meanwhile, the Cabot and the Newfoundland Steam Whaling companies embarked on their fifth and second seasons respectively. *Cabot* resumed the spring hunt off the south coast from Balaena[7] and caught eleven whales in Hermitage Bay in one week. At least 109 were taken by late October,[8] prompting observers to project even better prospects for the forthcoming season at Snook's Arm.[9] As a result, several families moved there from around Green Bay to find work.[10] The northern factory was now supplied by *Viking*, the company's second catcher (table 4.2) which had arrived in St John's from Norway on 9 July 1902 under the command of Captain R. Jacobsen (ill. 4.1).[11]

Hopes that the industry would be as productive as in previous years were diminished, however, when a St John's newspaper reported that "the whale

Table 4.1

Cape Broyle Whaling and Trading Co. Ltd of *Puma*

Incorporation date: 11 June 1902

Nominal capital: $60,000 ($100 shares); $120,000 ($100 shares), 21 October 1903.

Shareholder	*Address*	*Occupation*	*$100 Shares*
24 October 1902			
Bowring Bros.	St John's	merchants	200
Cabot Steam Whaling Co. Ltd	St John's	merchants	150
Michael P. Cashin	Cape Broyle	merchant	45
John S. Munn	St John's	accountant	40
(Plus 10 others with 1–6 shares, total 477)			
1 February 1904			
Bowring Bros.	St John's	merchants	468
John S. Munn	St John's	accountant	88
John Harvey	St John's	merchant	88
Kenneth R. Prowse	St John's	merchant	63
Michael P. Cashin	Cape Broyle	merchant	48
Edward J. Harvey	St John's	merchant	27
Adolph Nielsen (estate)			24
William V. Whiteway	St John's	barrister	20
(Plus 21 others with 2–18 shares, total 996)			

Later Holdings

6 March 1905; 28 February 1906: 1,200 (including Dr L. Rissmüller, 70; Estate Adolph Nielsen, 24).

Dissolution Date: 1 March 1906

Liquidation Accounts:

a) Debts: $47,498.79 ($47,435.64 owing to Bowring Bros.)

b) Sales: Factory, machinery, *Falcon*, at Cape Broyle	= $10,000
Factory, machinery, *Hawk*, at Cape Broyle	= $10,500
Total	= $20,500

c) Outstanding balance owing: $26,998.79

Source: Cape Broyle Whaling and Trading Co. Ltd, 1902–06.

factory at Snook's Arm will do nothing this year, for whales are very scarce north. One of the oldest settlers at Bett's Cove states that he has not seen a fish this year; an unprecedented thing in the bay. Some are of the belief that these fish are being killed out by the whaling steamers."[12]

Viking hunted "long distances" from Snook's Arm but caught only seventy-eight whales in her first northern season,[13] compared to 102 and eighty-six taken in each of the two previous years by Captain Bull in the *Cabot*. However, by using two catchers and stations the Cabot Steam Whaling Company was still able to capture 187 whales in 1902, their biggest catch to date. Since the new whaling regulations now restricted each company to one catcher,[14] they were allowed to use both vessels for this season only.[15]

The Newfoundland Steam Whaling Company also had a successful hunt off the south coast.[16] *Puma* (ill. 4.3) brought seventy-seven whales into Reuben's Cove before the season ended in early September. For the rest of the year the ship worked out of the company's second station, recently opened at Rose-au-Rue in Placentia Bay[17] (fig. 1.9) after it had been decided in late 1901 to expand operations into these previously unexploited waters. The new station employed 160 workers, including at least forty brought from St John's because of a local labour shortage.[18]

Whales were abundant in Placentia Bay, and *Puma* caught thirty-two in just five weeks.[19] Since the government now required whole-carcass utilization, guano plants were built at all stations. The factory at Rose-au-Rue was established by Dr Rissmüller and subsequently operated "full blast," as the *Evening Herald* reported: "the machinery is working splendidly, 60 men are employed there and in one week recently the carcasses of 14 whales were converted into fertilizing material of the best quality, shipment of which will soon be sent abroad."[20] The season ended in late December, and a slipway was built to hold *Puma* until spring.[21] The 1902 season was the ship's best to date, providing 155 whales for Rose-au-Rue and Reuben's Cove.[22]

Table 4.2
Registry Specifications of *Viking*

Official No.	*115467*	
Registration	St John's, 1902	
Built	Christiania, Norway, 1898	
Construction	Steel	
Dimensions	Length	88'
	Width	17'
	Depth (hold)	10'
Gross Tonnage	113	
Horsepower	35	

Source: *Mercantile List and Maritime Directory*, 1904.

Ill. 4.1 Captain R. Jacobsen and Newfoundland wife, Josephine White, Ferryland, c. 1903 (C.A. Coffey)

Despite its operational success, the company experienced some difficulties, including a particularly horrible accident at Reuben's Cove in which a worker died after falling into "a huge cauldron filled with acid."[23] Shareholders were also subjected to two court actions: Peter Hayes of Reuben's Cove unsuccessfully argued that he had been fraudulently induced to sell his land, and the Cabot Steam Whaling Company sued for alleged infringement of harpoon patent rights held by them.[24]

More entrepreneurs recognized the possible financial rewards from this new industry, and foreign involvement increased with the arrival in 1901 of Anders Ellefsen (ill. 4.2), a member of a prominent Norwegian whaling family. The government welcomed foreigners willing to set up businesses and gave

Ill. 4.2 Anders Ellefsen and family, Aquaforte, 1908 (Ellefsen Papers)

Ill. 4.3 *Puma* at Reuben's Cove, c. 1902 (Smithsonian Institution)

Ellefsen "full rights as to a British subject to establish himself,"[25] thus circumventing the 1902 act that allowed only British subjects to be employed on the whaling stations.

Ellefsen bought land at Aquaforte on the east coast of the Avalon Peninsula[26] (fig. 1.9) and shipped in the first construction materials during February 1902 on the schooner *Fashoda*. The station was projected to be completed by 25 May 1902[27] (ill. 4.4), and in April the *Evening Herald* reported that the buildings were "large and commodious. The slip for receiving the fish is almost completed and about 50 men have been working there the past few months. Thirty permanent hands will shortly be shipped and the new steam whaler will be here by May 20."[28]

Ill. 4.4 Aquaforte Whaling Station, c. 1904 (Ellefsen Papers)

The company's catcher *Humber* (ill.4.5) arrived in St John's from Norway on 6 June 1902. Unlike others in Newfoundland waters, the vessel remained registered in Norway, since the new company was purely a Norwegian operation.[29] *Humber* was "a shade smaller than the other whale steamers, being sloop rigged, and 22 tons. She is very fast and capable of doing 12 knots in smooth water."[30]

Humber began hunting on 17 June under the command of Andreas Ellefsen, Anders's father.[31] Twelve whales were taken by 27 June, and the carcasses were moored to be processed after the station was completed.[32] Ninety-three whales were killed in less than three months, mostly between Cape Race and Bay Bulls (fig. 1.1), although some were taken off Cape Bonavista.[33] Catching ended in December with a first season's total of 108.[34]

The new operation received a great deal of attention from local newspapers. The *Evening Herald*, for example, related an incident in which "an enormous fish, apparently nothing less than 150 ft. long was harpooned. The fish rose like a small house out of the water, and rushed the steamer. It was necessary to use every precaution and manoeuver clear of the angry monster who seemed determined to do damage. All sorts of things were tried to secure it, but, finally fearing the steamer might receive damage in one of the mad rushes, the line was cut and the ship ran away from the fish."[35]

Like the three earlier companies, Ellefsen experienced start-up problems. Although the 1902 act required guano plants to be built at each station, this was not immediately done at Aquaforte. Residents therefore complained about environmental pollution: "Two whales are on the slip at Aquaforte and 46 to date have been put ashore at Ferryland. The people contend that the carcasses of whales should be towed away out to sea. Recently two drove in on the beach, tore up several [cod fish] traps and the people had to keep their windows down owing to the stench arising from them. The ship later towed them to sea."[36]

Ill. 4.5 *Humber*, Aquaforte, c. 1904 (Ellefsen Papers)

Table 4.3
Catch and Production, Newfoundland and Labrador Shore-Stations, 1901–02

Season	*Catch*	*Oil (Tons)*	*Barrels per whale*	*Oil ($)*	*Whalebone ($)*	*Total ($)*
1901	258	633.5	14.7	54,221	13,550	67,771
1901–02	450	1275.5	17.0	112,859	12,285	125,144

Source: Newfoundland, *Annual [Fisheries] Report for 1903*, 1904.

Catching at Aquaforte slowed by mid-September, and ended in December with a first season's total of 108.[37]

The 1902 season was the most productive to date for the growing industry: 450 whales were caught (*Cabot* 109, *Humber* 108, *Puma* 155, *Viking* 78),[38] oil yields improved, and product values doubled over the previous year (table 4.3). Those few shares in established companies that came onto the market attracted high prices. Four $100 shares in the Cabot Steam Whaling Company, for example, sold for $230 each and eight in the Newfoundland Steam Whaling Company for $225.53 each.[39] The 1903 season was therefore expected to be equally productive.

Most of the oil continued to be exported to Scotland via St John's. *Virginia Lake* brought 1,009 barrels to St John's from Balaena and Reuben's Cove, and some arrived from Rose-au-Rue on the schooner *Rosebud.* The *Buenos Ayrean* took 746 barrels to Glasgow in June, followed by the *Livonian* with another 2,113 in July. The Norwegian vessel *Frenad* also left Aquaforte for Grangemouth in September with the first cargo of oil from this most recent addition to the industry.[40]

Local whaling was booming.[41] Hunting had extended to previously unexploited waters, and catches, yields, and exports had increased. Investor confidence was enhanced, and the government received forty-five additional applications to build new stations or renew existing licences between July 1902 and December 1903 (fig. 4.1, appendix 3). Further growth of the industry was inevitable.[42]

THE 1903 SEASON

The Newfoundland Steam Whaling Company

At the beginning of February 1903 the Newfoundland Steam Whaling Company sent *Puma* to Rose-au-Rue.[43] Whales were plentiful, and thirty had been captured by the end of March, mainly from Placentia and St Mary's bays.[44] This increased to seventy by the end of May, when the station was closed temporarily.[45] Reuben's Cove was then reopened for the summer fishery,[46] and *Puma* continued what was predicted to be "the best season since the industry started."[47] Whale stocks were still considered healthy on the south coast. It was reported that "the whaler *Puma*, some little time ago, passed thro a school of no less than 15 'sulphur- bottoms,' the largest kind of these fish, in Fortune Bay, showing that they are still plentiful despite the slaughter among them"[48] (fig. 1.1). An additional eighty-two whales were killed, and it was predicted that one hundred men would be employed if these catch rates could be maintained.[49] *Puma* returned to Rose-au-Rue at the end of July and hunted until mid-October,[50] when the whales were reported migrating out of coastal waters.

The Norwegian crew went home in November on *Carthaginian* after completing their three-year contract allowed under the 1902 act.[51] While *Puma* continued to hunt from Rose-au-Rue using local station workers, there is no record that the crew killed any whales. Nevertheless, their attempt brought about renewed demands on the government for Newfoundlanders to be trained, as specified in the act, for this more specialized and lucrative work.[52]

Again with a Norwegian crew, *Puma* took 260 whales from both of the company's stations in 1903, enabling them to declare a 40 per cent dividend and to transfer "a sum equal to another 20 per cent" to its reserve account.[53] Several shipments of products were made,[54] including bones for a guano factory established by Dr Rissmüller at Windsor, Nova Scotia.[55] Products were also sold on the Newfoundland market for the first time. The Colonial Manufacturing Company of St John's stationed a purchasing agent at Rose-au-Rue to buy guano to be sold locally as fertilizer, advertised as being better and cheaper than imports.[56]

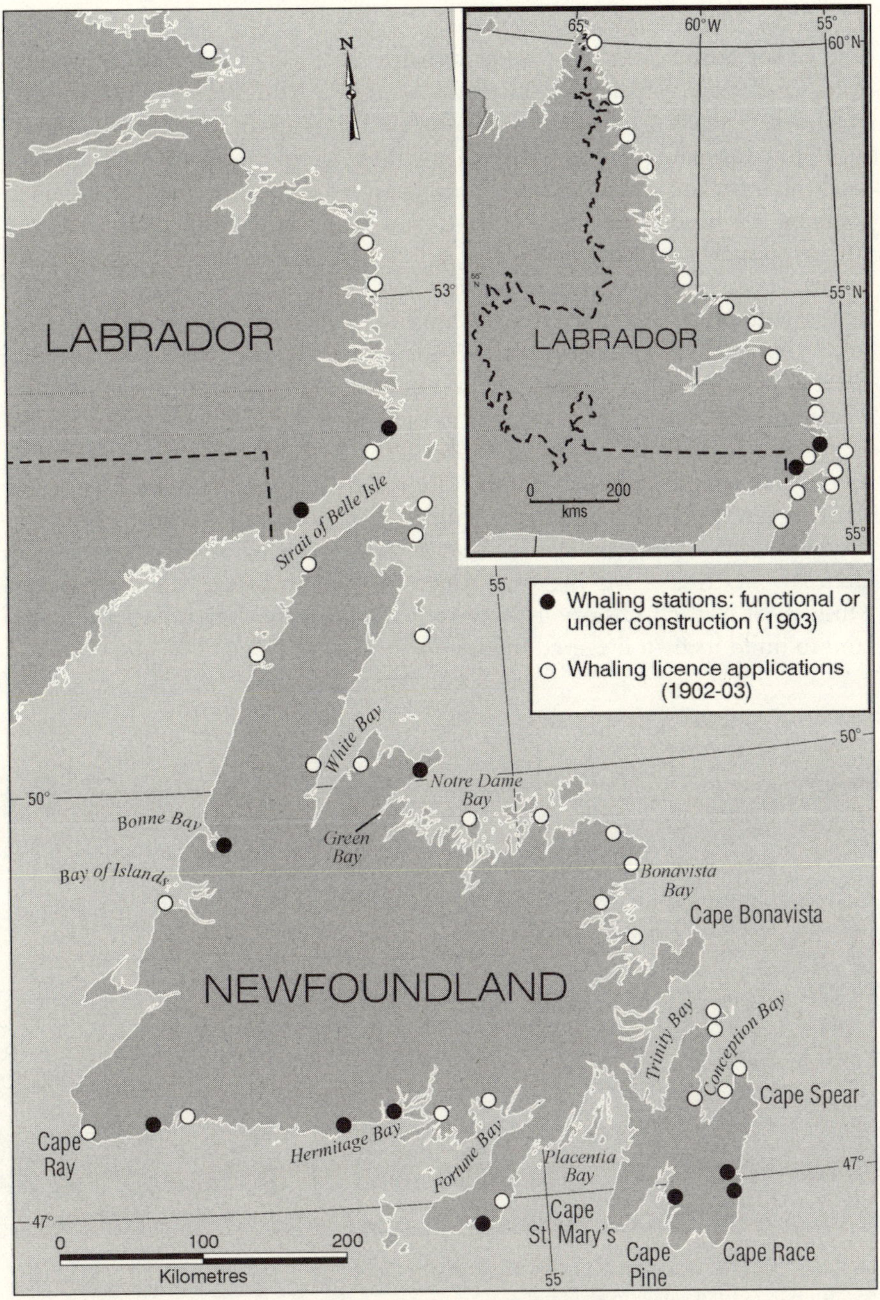

Fig. 4.1 Whaling station and licence application sites, 1902–03.

Source: Licence applications: Newfoundland, *Annual [Fisheries] Report for 1903*, 190; stations: various.

The Cabot Steam Whaling Company

The Cabot Steam Whaling Company began its season later than in previous years when poor weather prevented *Cabot* from leaving Balaena until the end of March. The crew killed 116 whales before bringing the vessel into St John's on 22 July for maintenance.[57] They then left for Snook's Arm and increased the total catch for the season to 211.[58] *Cabot* returned to St John's on 8 December, bringing the season to a close.[59] The oil and guano were again shipped primarily to Glasgow and Nova Scotia.[60]

Two events made 1903 noteworthy for the company. The founding shareholder, Adolph Nielsen, died from asthma on 2 November at the Crosbie Hotel in St John's.[61] That same year plaster casts of a male blue whale were made at Balaena by a team led by Dr Frederic Lucas of the Smithsonian Institution.[62] These and the skeleton were shipped to the United States[63] and used to build a life-sized model for the American exhibit at the 1904 St Louis Exposition.[64] The model was later transferred to the National Museum of Natural History in Washington, D.C.[65] Lucas left the Smithsonian to become curator-in-chief of the Museums of the Brooklyn Institute of Arts and Sciences in 1904, "the day after the skeleton of the big Blue whale was put in place."[66] The American Museum of Natural History in New York also used the data from Newfoundland to build its own life-sized model (ills. 4.6, 4.7).[67]

Left: Ill. 4.6 Blue whale model under construction, 1906
Right: Ill. 4.7 Blue whale model, 1907

Ill. 4.8. Cape Broyle station, c. 1906 (Ellefsen Papers)

Ill. 4.9 Postcard, Cape Broyle (G. Lilly)

The Cape Broyle Whaling and Trading Company

Construction of the station and guano factory at Cape Broyle (ills. 4.8, 4.9) resumed in January 1903,[68] and the company became important users of local labour. Thirty men from Toad's Cove (Tor's Cove), for example, began work on 1 March for more money than they could make from cod fishing.[69] People also arrived from St John's in search of work, and fifteen labourers who went on strike for higher wages were easily replaced.[70]

Viking under Captain Jacobsen left St John's on 25 February 1903 for Cape Broyle[71] and made its first whaling voyage in early March. Hunting was short lived, however. A harpoon line jammed in the winch, forcing the vessel to return to St John's for repairs.[72] Whaling resumed on 13 March, and Jacobsen killed a blue whale two days later, the first for the new company.[73] Eight whales were captured in the following month,[74] the catch totalling seventy-eight by early June.[75] As was common at other stations, flensing could not keep up with catching, and some carcasses were temporarily moored in adjacent Church

Cove.[76] Often hunting in company with *Humber* from nearby Aquaforte,[77] *Viking* brought ninety-eight carcasses back to Cape Broyle in this first year. *Retriever* (Captain Donnelly) took 180 tons of the company's oil to Glasgow in November 1903.[78] The smooth beginning to the Cape Broyle operation was unfortunately interrupted when the bunkhouse was destroyed by fire on 16 July.[79]

Aquaforte (Anders Ellefsen)

Anders Ellefsen experienced a similar calamity at Aquaforte when he and his new bride, Eda Melvin of St John's,[80] returned home in January 1903. Their house burned down after an oil lamp exploded, and the couple lost everything. Fortunately, prevailing winds prevented the fire spreading to the factory.[81] Despite the setback, *Humber* began to hunt and soon caught a female blue whale off Cape Spear in early February, valued at some $800.[82] By 14 May the vessel had captured thirty-two whales from "unusually large" numbers reported in east coast waters.[83] This kept the factory extremely busy, with "all it can do to cut up the carcasses and the prospects are exceptionally good for the company."[84] Because there was still no guano plant, it was again necessary for flensed carcasses to be towed out to sea and set adrift. Some were alleged to have drifted into Placentia Bay and damaged fishing gear, though an investigation by the recently appointed government whaling inspector, Charles Way, failed to substantiate the claim.[85] *Humber* killed 230 whales by the end of the season,[86] and *Gard* loaded 2,500 casks, "practically all the whale oil in the place," and some bone for Glasgow during the first week of October.[87]

Expansion

The directors of the Cape Broyle Whaling and Trading Company made a significant contribution to the further expansion of the industry when they decided to build the first "modern" whaling station on the Labrador coast.[88] The *Algerine*, skippered by Captain A. Jackman, took construction materials to Cape Charles (fig. 1.9) on 18 September 1903.[89] Although the factory had been almost completed by the beginning of October,[90] a catcher was not immediately available, so hunting could not start until the following summer.

Most residents living near the factories welcomed the growth of shore-station whaling because it opened new seasonal employment opportunities. People in St Mary's Bay, for example, petitioned the government to support the construction of a station by the St Mary's Whaling and Trading Company (table 4.4).[91] The company had been incorporated in St John's on 27 July 1903 with a nominal capital of $60,000. All 456 shares sold were held by Newfoundlanders, including Bowring Bros. and Michael P. Cashin, already principals in the Cape Broyle Whaling and Trading Company. Land was leased at Riverhead (fig. 1.9) for forty years from 15 August, and Cashin was issued a whaling licence on 26 August 1903.[92] The station was completed in December and included the "the factory [which] is 45 by 45 feet, 20 feet high and besides he has erected a new wharf, a bungalow, forge, coal shed of large capacity (filled with coal) after completion and a platform 2,000 feet long. The whole was done in 7 weeks, with 12 men at work. A new steamer is now in course of construction for the company at Norway and will be out here in April."[93]

The industry also expanded to the Burin Peninsula when Dr Rissmüller, who had initially come to Newfoundland for the sole purpose of building and operating guano plants, opened his own whaling station at Little St Lawrence (fig. 1.9, ill. 4.10). He had assistance from the Harvey family, already major investors in the Cabot Steam Whaling Company. Construction began in July 1903, and Captain T. Bonia, a member of the House of Assembly in the Newfoundland Legislature, was appointed as the first station manager.[94] He was described as "a hustler": "in his competent hands the undertaking will speedily be pushed to a successful completion."[95] Unfortunately, operations were temporarily suspended in early October for unidentified reasons, placing some fifty men out of work.[96] Nevertheless, the station continued to be seen locally as the solution to serious unemployment,[97] and an editorial in the *Evening Herald* commented that "the people, who did not do over well at the cod fishery the past summer, are delighted at this substantial adjunct being provided there."[98]

As others had done before, Rissmüller bought a catcher in Norway (table 4.5). *St Lawrence* (ill. 4.11) entered St John's harbour on 6 October,[99] was transferred to British registry, and arrived at Little St Lawrence on 15 October 1903.[100] The vessel began hunting on 1 November but could only take seven whales by the end of the season.[101]

A third station was built on the south coast of the island at Dublin Cove (fig. 1.9) by the Mic Mac Whaling Co. Ltd, incorporated on 29 September 1903 with a nominal capital of $150,000, the largest initial offering to date (table 4.6). Construction began in December and was expected to be finished by the

Table 4.4
Major Shareholders in the St Mary's Whaling and Trading Co. Ltd, 20 December 1903

Shareholder	*Occupation*	*$100 Shares*
Michael P. Cashin	merchant	55
Bowring Bros.	merchants	50
Tasker Cook	shopkeeper	50
Job Bros.	merchants	50
Patrick McGrath	editor	45
Martin Furlong	solicitor	20
Daniel Ryan	merchant	20
Frank Connors	saddler	20
Ludwig Rissmüller	chemist	10
Plus 24 others at 10 shares and below		
Total		456

Source: St Mary's Whaling and Trading Co. Ltd, 1903.

Ill. 4.10 Flensing at Little St Lawrence, c. 1906 (R. Street)

Table 4.5
Registry Specifications of *St Lawrence*

Official No.	*117295*	
Port of registration	St John's, 1903	
Built	Christiania, Norway, 1903	
Construction	Steel	
Dimensions	Length	93'
	Breadth	17'
	Depth	10'
Gross Tonnage	111	
Horsepower	50	
Owner	John Harvey, St John's	

Source: *Mercantile List and Maritime Directory*, 1904.

Ill. 4.11 *St Lawrence* with whale, British Columbia, c. 1907 (Vancouver Maritime Museum)

Table 4.6
Initial Shareholders of the Mic Mac Whaling Co. Ltd, 29 September 1903

Shareholder	*Occupation*	*$100 Shares*
William C Job	merchant	215
Robert B. Job	merchant	100
Marmaduke G. Winter	merchant	100
Kenneth R. Prowse	merchant	100
Colin Campbell	merchant	75
Alexander McDougal	merchant	50
Hon. James S. Pitts	merchant	50

Source: Mic Mac Whaling Co. Ltd, 1903.

Table 4.7

Major initial Shareholders in the Labrador Whaling and Manufacturing Co. Ltd, July 1904

Shareholder	*Occupation*	*$100 Shares*
W.D. Reid	railway contractor	150
Daniel A Ryan	merchant	50
James Ryan	merchant	50
Charles Dawe	merchant	50
and 31 others at 30 shares or less		635

Source: Labrador Whaling and Manufacturing Co. Ltd, 1904.

Table 4.8

Initial Shareholders in the Harbour Grace Whaling Co. Ltd, 9 August 1904

Shareholder	*Occupation*	*$100 Shares*
W.D. Reid	railway contractor	50
Dr L. Rissmüller	manufacturer	50
J.B. Ayre	gentleman	30
Charles Dawe	master mariner	20
James Knight	merchant	20
S. Blandford	gentleman	10
R.B. McRae	merchant	5

Source: Harbour Grace Whaling Co. Ltd, 1904.

Table 4.9

1903 Whaling Season Production, Newfoundland and Labrador

Items	*Tons*	*Value ($)*
Guano	585	–
Whale bone	773	9,950
Finners	41	–
Oil	2,664	256,372
Total	–	266,322

Source: Newfoundland, *Annual [Fisheries] Report for 1903*, 1904.

end of January in time for the 1904 whaling season.[102] The company also expanded the industry in Labrador by building a station at L'Anse-au-Loup (fig. 1.9). The factory was projected to be "a fine structure 53 by 52 feet, 32 feet high, and will be rushed to completion in the spring. Dr Rissmüller will also likely start to erect a [guano] factory there early next summer."[103]

A third Labrador station was also proposed for Hawke Harbour (fig. 1.9) by Daniel Ryan, a merchant from King's Cove, Bonavista Bay. However, the Labrador Whaling and Manufacturing Co. Ltd, incorporated on 7 July 1904 with a nominal capital of $100,000[104] (table 4.7), did not begin construction until 1905.

The Atlantic Whaling and Manufacturing Co. Ltd, incorporated on 8 February 1904 with a nominal capital of $250,000, extended the industry into Trinity Bay. Dr Rissmüller and Job Bros., with three hundred and two hundred shares respectively, were the major investors.[105] Again, residents adjacent to the proposed station at Maggotty Cove near Trinity welcomed the new employment possibilities.[106]

Two stations were also suggested for Conception Bay, but the one proposed for Portugal Cove–Bell Island by Joseph Roper did not materialize.[107] Construction on the other started at Ship's Head, Harbour Grace, where "about an acre in area ... [was] devoted to the building of the factory, store, outbuildings, etc., the whole to cost about $50,000."[108] This station was owned by the Harbour Grace Whaling Co. Ltd, incorporated on 9 August 1904 with a nominal capital of $100,000. However, only 185 shares were purchased in the original offer[109] (table 4.8). The company remained periodically active for many years, largely as a mooring and winter overhaul base for catchers from Labrador after 1920.

In late 1903 the successes of the Cabot and Newfoundland Steam Whaling companies were evident as local investors continued to be convinced that substantial profits could be made from the new industry. The construction of additional stations and catchers had brought the total annual catch to 1,308 whales during the 1902–03 seasons. Product values increased to $266,322, again more than doubling the net worth of the venture (tables 4.3, 4.9). Some $550,000 of private capital was now invested, and the government received $19,500 in licence fees.[110]

Despite the dramatic growth in seasonal employment, some participants now began to express fears that if the pace of expansion continued, "the fishery would, inside of twenty years be a thing of the past."[111] For the first time, press reports also speculated on the sustainability of the industry, since although "whaling stock is regarded as a good investment ... whether it will continue so admits of some doubt."[112] The potential for decline was further aggravated by ongoing poor hunting practices. Crews sometimes only wounded or failed to secure their catch, for example, and drifting carcasses were often reported.[113] Of greater significance was the fact that 44 per cent of the catch was female. Little attention was paid to these concerns, however, and expansion continued into the 1904 and 1905 seasons.

CHAPTER 5

The Peak, 1904

The rapid growth of the poorly regulated industry and the overexploitation of stocks made the Newfoundland Department of Fisheries advise against further expansion just five years after the first whale was killed in June 1898.[1] The pace at which the initial phase of whaling had developed required more effective management, they argued, since in "Norway, owing to unrestricted killing, the whale fishery was depleted in fourteen years. This experience should be a guiding light to Newfoundland in the path of prudence."[2]

The warning was not heeded. The *Evening Herald* reported in early 1904: "Another new whaling company is … in the course of formation, all the principals being young men of this city. They are endeavouring to obtain a site for the factory on the west coast, and if this can be obtained the work of promotion of the company with other incidentals will be pushed quickly forward."[3] This continued interest was stimulated by the profits being earned by existing companies. The Newfoundland Steam Whaling Company, for example, had most recently declared a 40 per cent dividend for the 1903 season and transferred 20 per cent to its reserves.[4]

The government now recognized that its whaling legislation would have to be more stringently enforced if the industry were to be sustained. As Governor Cavendish Boyle pointed out in the 1904 Throne Speech, "The whale fishery has rapidly become a most important branch of our commerce, and during the present year a very large amount of additional capital will be invested therein. The value of this fishery, and the revenue that has already resulted therefrom, emphasize the wisdom of rigidly enforcing the law so widely adopted by this Legislature two years ago for its protection."[5]

The House of Assembly gave unanimous support. As one minister stated, whaling had become a significant revenue generator for Newfoundland: "This industry … under present methods, is a comparatively new one, and the returns from the money invested … have in some cases been phenomenal, while the enterprise has been the means of largely augmenting our revenues. There is every indication and every assurance at present that the coming year will see the whale fishery fleet largely increased, which will not only have the effect of giving additional employment to our people, but must necessarily swell the exchequer of our colony."[6]

The government also decided to review the effect of whaling on other fisheries before issuing more licences.[7] This was in response to continued criticism from fishermen who thought that recent catch reductions in species such as capelin were due to "the killing out of whales"[8] believed to drive the fish inshore when feeding.[9] In reality, it appears that whales were already being overexploited. Fishermen living near the station at Rose-au-Rue, for example, reported that whereas "they used to have to steer clear of whales, now one is very seldom seen."[10]

The whaling companies paid little notice to these concerns, especially since an increasingly profitable market for oil and whalebone was available to them in Dundee.[11] Baleen was in particular demand. With 2.25 tons being sold there for $33,750 in early 1904,[12] and only four tons reputedly left unsold on the global market, there were "no 'corners' in this particular commodity." Demand had outrun supply so rapidly that in seventy years the price of whalebone had risen from thirteen cents to nearly $7 a pound. "Yet never in the greatest years of the whale fishery did any vessel make so many captures as were reported by the Newfoundland whalers in the past season. An average catch of somewhat more than two hundred whales apiece was certainly a remarkable achievement for four vessels."[13] Optimism was translated into practice, with even greater catches and product values (table 5.1).

The production of fertilizer from ground bone and carcass residues increased after Job Bros., managers of the Atlantic Manufacturing Company in St John's, built a processing plant for Dr Rissmüller next to the whaling station at Aquaforte (fig. 5 1). The factory also used flensed carcasses bought from the St Mary's Whaling and Trading Company station at Riverhead. A good export market existed, and the *Regulus* was kept busy collecting "all the guano and whale material ready for shipment" from the stations at St Mary's, Cape Broyle, Aquaforte, and Trinity for transport to Charleston, South Carolina.[14]

In an attempt to increase profits, the industry also experimented with canning whale meat for human consumption. The flesh that in the past had been thrown away was "said to be a choice article when properly put up and a ready market is obtained for it in certain places. This will be another source of revenue for those interested and it is hoped the trial to be made will result satisfactorily."[15] A consignment of smoked and dried whale meat exported to New York was said to be "for all the world undistinguishable except on close

Table 5.1
Whale Oil and Bone Production and Value, Newfoundland Shore-Stations, 1897–98 to 1902–03

Season	*Oil Tons*	*Oil $*	*Oil $/Ton*	*Bone $*	*Total $*
1897–98	27	1,330	49.3	241	1,571
1898–99	211	14,439	68.4	1,167	15,606
1899–00	430	34,694	80.7	1,824	36,518
1900–01	665.25	54,221	81.5	13,550	67,771
1901–02	1,275.50	112,859	88.5	12,285	125,144
1902–03	2,664	256,372	96.3	36,761	293,133
Total	5,272.75	473,915	89.9	65,828	539,743

Sources: *EH*, 14 March, 12 November 1904.
Note: Slightly different numbers are provided in the two sources.

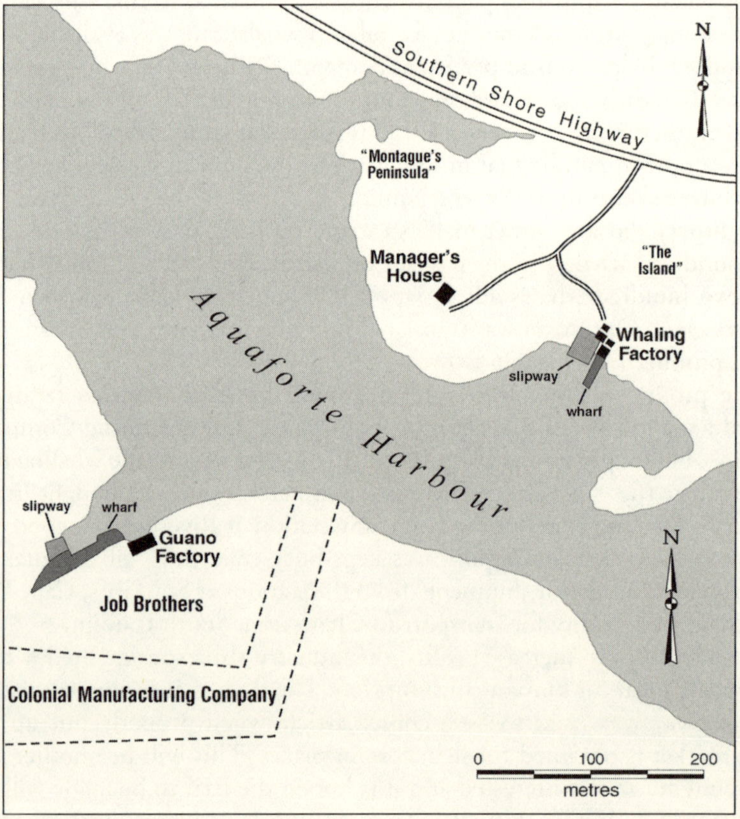

Fig. 5.1 Location of Aquaforte whaling station and guano plant, c. 1904.
Source: Dickinson and Sanger 2001

inspection from jerked beef. We know of no reason why a market would not be created for this excellent dietary. The whale is the cleanest feeding mammal alive and the meat is every bit as fine in texture and quite as palatable as many grades of ox beef. It can be sold at an infinitesimal fraction of the cost of the latter."[16]

As markets grew, more applications were made for whaling licences.[17] A St John's solicitor, James McNeily, proposed a station for Hare's Harbour, Fortune Bay.[18] Charles Smith of St John's, a "whale factory engineer and superintendent," considered building one at Little Cat Arm, White Bay.[19] Although neither was developed, the newly incorporated Harbour Grace Whaling Company called tenders for construction of its station on 19 July 1904.[20]

To increase the participation of Newfoundlanders in the more lucrative aspects of the industry, the House of Assembly proposed to ban the use of Norwegian catcher crews by any company for more than three years. The minister of Justice, however, considered this move unnecessary as the 1902 whaling act already permitted fines of $50 per day to be levied for infractions of this nature. Proposals to encourage the local construction of whale catchers by refunding duties on materials and increasing shipbuilding bounties were also defeated.[21] The industry thus continued to depend significantly on Norwegian expertise.

In sharp contrast to the situation in Newfoundland, coastal whaling in Norway remained in decline. Stocks were overexploited, and many complaints were received by the government about other fisheries being seriously damaged by the industry.[22] Whereas Norwegian hunting had from 1 February 1904 been restricted to specified coastal waters,[23] local whaling was now occurring virtually everywhere around Newfoundland and southern Labrador. The only exceptions were parts of the west coast and Great Northern Peninsula, or the "French shore." "Along the whole coast line of 800 miles [between Cape John and Cape Ray, fig. 1.1]… the whaling industry could not be established there (by the British), because the French contended it would interfere with their cod fishery, and for the sake of peace, England acquiesced," observed the *Evening Herald*. Pressure, however, was building to include these regions as well: the industry would "immediately be enterprised, for licenses will [be issued] … to applicants forthwith, and four whale factories will be erected this season on that shore."[24]

The first application to build a whaling station on this previously restricted coast was made on 25 April 1904 by Joseph Moore of St John's for a site at Port Saunders (Hawke's Bay). It was followed by a similar application on 4 June from Edmund J. Ryan, King's Cove, Bonavista Bay, for a station at Conche.[25] In reviewing these proposals the government estimated that the seventeen stations projected to operate in 1905 would yield products valued at some $700,000, more than double that earned in 1902–03 (table 5.1).[26]

This growth not only benefited company shareholders and employees but also generated significant work for construction and engineering companies in St John's. At the beginning of the 1904 season, the Avalon Steam Cooperage Company, for example, was daily producing over three hundred oil barrels of "exceptionally good quality, [and] all the whale and other oil factories are anxious to get their stocks as quickly as possible."[27] Similarly, the Reid Newfoundland Company was "very much rushed just now making boilers for the Atlantic Whaling Company's factories at Trinity and Safe Harbour, Bonavista Bay" with the contract "to furnish all the engines, tanks, winches and pipes to be used in connection with the operation of these factories."[28]

Newfoundland investors also attempted to increase their participation in the fledgling whaling industry in British Columbia, where "labour is cheap and buildings can be put up at about one quarter the cost of a factory here." Although their men, vessels, and equipment were eventually used, Newfoundland companies were unable to build stations in British Columbia since legislation prohibited construction by anyone with less than five years' residence.[29]

THE 1904 SEASON

Despite a slow start to the season because of poor weather and heavy ice,[30] record numbers of whales were caught (table 5.2) in 1904. Unfortunately, this success did not translate into increased sales and profits due to "a heavy glut" in the whale-oil market in Britain, "no fewer than 50,000 casks of that commodity from this island being still unsold … and prospects of disposal at a good figure being quite unsatisfactory. There will not be large dividends in the whaling business this year."[31]

Nevertheless, the market for baleen (whalebone) still appeared to be good, since the entire Dundee Arctic whaling fleet only killed twenty-two whales in 1902–03 for some sixteen and a half tons.[32] The Dundee vessels also returned at the end of 1904 with just "eight fish, representing among other things, about five and a half tons of bone. Quoted at $2,200 per ton, and rising,"[33] the high value baleen made a significant contribution to the Newfoundland industry's continued growth.

Snook's Arm and Balaena (Cabot Steam Whaling Company)

The first reported activity of 1904 for the Cabot Steam Whaling Company was in March, when *Fashoda* left St John's for Halifax with forty-five tons of bone produced during the previous season at Snook's Arm. Hunting started slowly at Balaena, with *Cabot*'s catch of eighteen being the lowest for the fleet at the end of May. This number, however, included a ninety-five-foot blue whale, "one of the largest" ever brought into the station. Conditions improved in June, with the vessel doing "good work," and a cargo of oil was shipped to Glasgow on *Flora* in July.[34] The station closed in early September after processing eighty-two whales (table 5.2). *Cabot* then moved north to Snook's Arm and took an additional fifty-nine by the time *Miss Morris* arrived from St John's in mid-November to load oil for Europe. The company's 1904 season appears to have been fairly successful, with factory workers "having made good wages."[35]

Table 5.2
Catch and Production, Newfoundland and Labrador Shore-Stations, 1904

Station	*Catch*	*Species*	*Oil (Gals.)*	*Guano (Tons)*	*Bone (Tons)*
Balaena	82	57b, 9h, 11f, 5s	108,435	290	423
Snook's Arm	59	4h, 55f	57,515	138	160
Reuben's Cove	142	82b, 7h, 23f, 30s	156,100	401	324
Rose-au-Rue	155	48b, 69f, 35h, 3s	224,560	489	387
Beaverton	28	22f, 6h	37,690	57.5	50
Little St Lawrence	112	37b, 65f, 9h, 1s	192,024	502	200
Cape Broyle	150	6b, 91f, 53h	122,194	411	578
Cape Charles	104	3b, 70f, 31h	90,594	193.75	190
Dublin Cove	6	4b, 2h	6,693	4.5	5
L'Anse au Loup	49	1b, 39f, 9h	53,544	51.25	90
St Mary's	63	12b, 31f, 20h	58,590	112	100
Aquaforte	198	9b, 103f, 86h	170,821	286	160
Trinity	68	3b, 59f, 5h, 1sp	115,416	294	126
Safe Harbour	59	2b, 52f, 5h	98,280	281	110
Totals	1,275	264b, 281h, 690f, 39s, 1sp	1,492,456	3,511	2,903

Source: Newfoundland, *Annual [Fisheries] Report for 1904*, 1905.
Species: b = blue, f = fin, h = humpback, s = sei, sp = sperm.

Ruben's Cove, Rose-au-Rue, and Beaverton (Newfoundland Steam Whaling Company)

Puma began hunting at Rose-au-Rue in late April and was doing "good work at whaling" during the first week of May. Later in the month *Alpha* arrived to load three hundred tons of oil stored from the previous season and ship it to Glasgow. By 18 June *Puma* had killed sixty whales, including several blues, and another oil cargo was loaded on *Electra* for the United Kingdom. *Puma*'s catch increased to ninety-three by 12 July, including forty-one blue whales. By the time the station closed in September, "whales being scarce," 155 had been processed. Meanwhile, *Lynx,* the company's other catcher, had similar success supplying the station at Reuben's Cove, it being "all the employees can do to keep one fish cut up before another arrives. They are short of men, as many have gone home sick having been unable to stand the smell from the guano factory."[36]

Ninety-eight whales were killed by the beginning of August, and "scores of whales were seen including a school of small ones." The catch increased to 113 by mid-August, "the best season's work since the industry started," and reached a total of 142 by the time the station closed in late September. *Lynx* then moved north to Beaverton, capturing an additional twenty-eight whales by mid-November when operations ended. The company shipped several cargoes, including guano on *Aureola* to Charleston, South Carolina; fertilizer on *Anver* to Norfolk, Virginia; bone on *Alberta* to Nova Scotia; and oil on *Morning Star* and *Electra* to the United Kingdom.[37]

Aquaforte (Anders Ellefsen)

Humber took only five whales in January because of severe weather and was then placed on dry-dock in mid-March for maintenance. On resuming hunting, the vessel caught eighteen whales by the beginning of May and led the fleet at the end of the month with thirty-six. The catch increased to over sixty by mid-June, and 183 in September. The station thus had "every prospect of going well over the 200 mark before the work finishes." When whales became scarce off the east coast of the Avalon Peninsula, however, *Humber* was forced to "travel north of Baccalieu in quest of them." By the time the season ended in early November, 198 whales had been captured.[38]

Cape Broyle and Cape Charles (Cape Broyle Whaling and Trading Company)

The Cape Broyle Whaling and Trading Company operated both the Cape Broyle and Cape Charles stations in 1904. The catcher *Viking*, purchased by majority shareholders Bowring Bros. from the Cabot Steam Whaling Company, was renamed *Falcon* to distinguish her from a famous Newfoundland sealing vessel of the same name.[39] Captain Jacobsen took his first whale, a humpback, in early April off Cape Broyle. By the beginning of November, he had killed 170, "and the prospects are bright for more." The catch included a female blue with a seventeen-foot embryo, "the largest ever taken from any whale yet captured there."[40] As whales migrated beyond the reach of the Avalon Peninsula stations, *Falcon* sailed for the more northerly hunting grounds and took at least another four from Trinity Bay and two off Baccalieu Island. Oil and bone from Cape Broyle were shipped on *Dunure* to Glasgow, and bone went to Nova Scotia on the schooner *Fauna*.[41]

The company also extended its operations to Labrador, sending men and machinery there on *Aurora* in late April to finish work on the station at Cape Charles.[42] Their second catcher, *Hawk*, left Christiania (Oslo) on 20 April and arrived at Cape Charles, via Greenspond and St John's, on 8 June. Whales "were plentiful before the ice came in and all are anxious for the start."[43] Only nine whales were caught by early July, however, due to heavy ice. To add to the company's problems, local residents complained that the 1902 whale fishery act was being contravened when offal was allowed to "float about the bay and has been seen down to Gull Is., near Battle Hr. The stench is abominable and traps will be spoiled if this stuff is allowed to float about indiscriminately."[44] Whales became plentiful in July, with eighty-two killed by mid-September "and prospects … good for soon reaching the century mark."[45] *Hawk* returned to St John's on 12 November after taking "104 fish, the last being a sulphur bottom, 98 feet long, one of the largest ever caught on the coast. The weather has been extra stormy of late and as there was work to be done on the factory operations were suspended. On the way south near Belle Is. the captain reports passing school upon school of whales, some enormous fish among them, and in all he reckoned about 400 fish."[46] Altogether, approximately eighty station-workers received an average seasonal wage of $100 "at $22 per month and 9 cents overtime, of which there was much to be had."[47]

Riverhead, St Mary's Bay (St Mary's Whaling and Trading Company)
The new catcher *Avalon* arrived in Newfoundland from Norway in mid-May and made a "fine opening of the season," killing two blues and one humpback in its first two days of hunting. The capture of two additional blue whales the following day elicited the comment that "if she keeps up this record she will soon block the operations of cutting up at the factory." By the beginning of June, the factory was "in full rush, the whaler *Avalon* having taken 5 fish, 2 sulphur bottoms and 3 finbacks."[48] At the end of the month, the station was "in full swing and all the operatives there are busy under supervision of Mr. Lewis Mullowney. The whaler *Avalon* since starting has taken 34 fish, capturing 6 Sunday and 2 Tuesday. One big fellow was a sulphur bottom 86 feet long. These are mostly finbacks and were captured in Placentia and St. Mary's bays, while 6 were taken on the southern shore. This is a good start."[49]

Two months later *Avalon's* catch had increased to forty-five, and to sixty-one by the beginning of November, "the greater part of which were taken 50 to 100 miles off Cape St. Mary's." By the middle of the month, *Avalon* was reported to be in St John's with whale oil, thus bringing the station's first season to a close.[50]

Trinity and Safe Harbour (Atlantic Whaling and Manufacturing Company)
In June 1904 the schooner *Hydrangea* brought machinery made by the Reid Newfoundland Company in St John's to the new station at Trinity (ill. 5.1), and two nine-ton boilers arrived by rail and sea.[51] The catcher *Fin* under Captain Nils Neilssen entered St John's on 11 July from Christiania (Oslo), and began hunting from Trinity on 19 July. The catcher took a fin whale on the maiden voyage, and the industry's first sperm whale six days later, "a most valuable fish." Although whales were plentiful in Trinity and Bonavista Bays, *Fin* could only take thirty-five by 9 September because of "stormy weather." Conditions improved thereafter, and a total of sixty-eight were caught before *Fin* went on to dry-dock in St John's in late November for an end-of-season overhaul before returning to Trinity for winter lay-up. Oil and guano were taken to St John's on the schooner *Pansy*.[52]

Construction of the company's other station at Safe Harbour (ill. 5.2) also began in the spring, and the newly built catcher *Hump* (ill. 5.3) arrived in St

John's on 15 July from Christiania via Galway. Hunting began on 19 August, with two whales being killed on the first day. Whales continued to be "very plentiful in Bonavista Bay," allowing *Hump* to average one per day. The catch increased to over thirty by mid-October. *Hump* returned to St John's prior to over-wintering at Safe Harbour. Oil was shipped to Europe on *Ellen Lloyd*.[53]

Dublin Cove and L'Anse au Loup (Mic Mac Whaling Company)

The Mic Mac Whaling Company began its first season using both of its newly completed stations. *Mic Mac* (ill. 5.4) sailed from St John's to Dublin Cove on 5 May but ran aground entering Port-aux-Basques, "a serious matter for the new whaling factory [which] … will materially hamper operations there." Refloated the next day, *Mic Mac* returned to St John's where damaged plates were replaced and the keel straightened. It left the dry-dock on 11 June and returned to Dublin Cove three days later. Whaling began in mid-June but with little success.[54]

Construction continued at L'Anse au Loup, and *Mic Mac* moved north to Labrador in August. There results were better, and by early September the catcher was averaging six whales per trip and "the hands in the factory are working overtime." Twenty-seven were caught by mid-October, although at some cost, since the *Mic Mac* "landed four of her crew, suffering from slow poisoning, contracted in some mysterious manner, and one of them has just died in Battle [Harbour] hospital." *Mic Mac* left Labrador on 10 December for Dublin Cove only to run aground again, this time in Lark Harbour (fig. 1.1) during a snow storm. Oil and guano were shipped from L'Anse au Loup on the *Nimrod*.[55]

Little St Lawrence (Dr Ludwig Rissmüller)

A legal case involving the Newfoundland Steam Whaling Company, operators of the station at Rose-au-Rue, and the Newfoundland government regarding the location of a new station proposed by Dr Rissmüller for Little St Lawrence was argued before the Privy Council in London on 31 April. The council ruled in favour of the government and Rissmüller.[56] The temporary twelve-month licence issued in October 1903 by the Newfoundland Supreme Court was immediately made permanent. Although *St Lawrence* did not begin hunting until late in 1903, the quick capture of seven whales suggested that there would be "much labour for the residents."[57]

The station continued to provide good employment opportunities for nearby communities at the beginning of the 1904 season. Sixty-three whales were taken by mid-September, including "a sulphur [blue] 80 feet long, the largest fish taken by her." At least 112 were eventually processed (table 5.2) in this first full season, with oil and other "products" shipped to England on *Viola*, *Carpasian*, and *Regulus*. These included glue, canned meat, and sausages developed by Rissmüller as "an experiment that is thought will become a marketable

Ill. 5.1 Trinity (Magotty Cove) whaling station, c. 1906 (G. Lilly)

Ill. 5.2 Whale on slipway, Safe Harbour, c. 1907 (G. Lilly)

Ill. 5.3 *Hump*, Rose Blanche, c. 1910 (E. Kiserud)

Ill. 5.4 *Mic Mac*, renamed *Sebastian* (1907), then *Saanich* (1916) (Vancouver Maritime Museum)

commodity in time. Parties at the factory prefer the whale steak to any other kind of food, and the sausage is said to be exceptionally good and hardly distinguishable from beef."[58]

The industry reached its peak in 1904, expanding from one catcher and station processing ninety-one whales in 1898 to ten vessels operating from fourteen stations just six years later. In 1904 1,275 whales yielded 1,492,456 gallons of oil, 2,903 tons of whalebone, and 3,511 tons of guano. This success again convinced additional investors that substantial profits were still possible, and expansion continued. Unfortunately, it was becoming increasingly evident that stocks could not sustain the growing exploitation levels. Nevertheless, pre-season cautions issued by the Department of Fisheries about over-hunting were ignored, and the industry entered the end-phase of its first cycle.[59]

CHAPTER 6

The First Decline, 1905–17

The 1905 season saw catches decline for the first time. Three new factories had opened, and their catchers *Cachalot* (Hawke Harbour), *Baccalieu* (Harbour Grace), and *Port Saunders* (Hawke's Bay) increased the fleet to a high of fourteen (fig. 6.1).[1] A station proposed for Lark Harbour by the Neptune Steam Whaling Co. Ltd, to be supplied by the catcher *Neptune*, did not open.[2]

As in previous years, whales appeared plentiful at the beginning of the season.[3] Anders Ellefsen was thus able to write from Aquaforte to Ragnvald Berg, one of his Norwegian financiers, "I should imagine that the first thing to start with is how the hunt is going, we had 4 blues in 3 days, that looks nice doesn't it? Henrik [Anders's brother and captain of the Aquaforte station's catcher, *Humber*] reports many whales."[4] Animals were "visible everywhere," and *Humber* had excellent numbers into May.[5] While this initially seemed to indicate that the catches made from Aquaforte in 1902–04 would be sustained, by the end of the month *Humber* had only taken 12 per cent of the 226 whales killed by the eight catchers then operating "on the south coast."[6] This situation deteriorated even further as the season progressed, and the masters reported at the end of June that they had not caught "a half dozen between them."[7] By early July only forty-six whales had been processed at Aquaforte, about 50 per cent of the previous season's productivity for the same period.

Other companies fared no better.[8] Although there was some later improvement as the fleet followed the migrating whales north, the 1905 season was regarded by all as a failure. A report in the *Evening Herald* described the perilous state of the industry: "There is much talk about town these days respecting our whaling industry and the ill-fortune of most of the companies now operating, with the prospect of an early closing of some, never to re-open as at present constituted. Enormous sums of money have been sunk in them – with no hope of realization ... only two or three concerns made any money this year, and these not very much, so that their profit is likely to be placed to reserve instead of declaring a dividend."[9]

The 1905 catch declined more than 30 per cent over the previous season, despite the deployment of four new vessels (fig. 6.1, table 6.1). The Newfoundland Department of Fisheries considered this "a very large deficiency ... [that] demonstrates the great scarcity during the past year."[10] Increased global catches also over-supplied oil and bone markets and caused prices to decline even further. Grade 1 oil fell from £21 a ton in 1901 to £16 in 1904, and several Newfoundland companies were forced to sell at £14 in 1905.[11] This had serious implications for the future of the industry since "the price of whale oil has declined one-half in three years and, consequently, it requires twice as many fish to make a factory pay now as it did then."[12]

Investors saw their profits declining and petitioned the government to replace the standard licence fee with one based "on the profits of each concern, rather than in a lump sum of $1,500 as at present."[13] They also asked that the $50 per ton coal tax imposed on each station be removed.[14] Neither petition was successful,[15] contributing to the further decline of the industry. The Atlantic, Mic Mac, and St Mary's whaling companies were the first to be liquidated, the sale of their factories, equipment, and vessels advertised in local newspapers.[16] There was little interest, and the three companies were instead restructured as the Phoenix Whaling Co. Ltd, with Job Bros. of St John's as managers and principal investors.[17]

The problems facing the industry were well described in the annual report of the Newfoundland Department of Fisheries for 1905:

> The price of produce was again low during the year ... so that the few companies which were successful as regards the catch scarcely paid running expenses. The catch of most of the companies was a comparative failure, and the result in several cases has been the winding up of the companies, with the loss of their entire paid up capital ... The expenses of carrying on this business are very heavy, and like all other fisheries the results are very fluctuating and uncertain.[18]

The disastrous season undermined the confidence of investors, and the end of the industry was predicted.[19] Low catches in the early part of 1906 (table 6.2) raised more concerns about the industry's future, even though four stations had been closed.[20] The rapid deterioration[21] was for the first time specifically blamed on stock depletion brought on by over-hunting.[22]

The difficulties facing the local industry are best illustrated by the Aquaforte whaling operation (see chapter 7). With catches declining, Anders Ellefsen asked the Colonial Cordage Company in St John's for additional time to pay for a shipment of hemp lines for *Humber.* Recognizing the problems confronting the whaling companies, Colonial Cordage gave him four months credit in the hope that catches would improve.[23] Although layoffs were occurring in associated industries,[24] there was still sympathy for the whaling companies. The

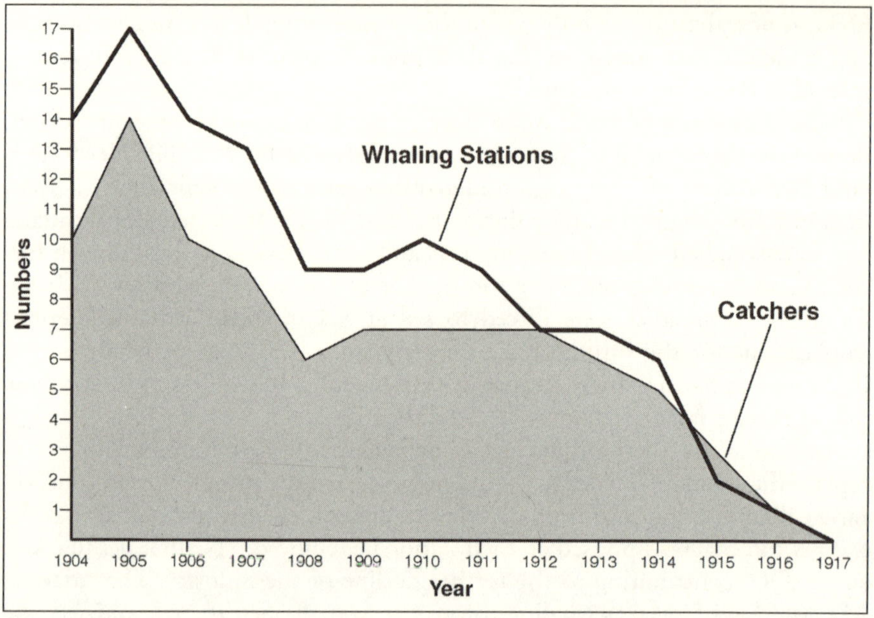

Fig. 6.1 Newfoundland and Labrador whaling stations and catchers, 1904–17.

cancellation of another order from Ellefsen,[25] for example, brought this response from Colonial Cordage: "We are extremely sorry to hear that your heavy investment in plant … is not likely to be remunerative this year. We hope that the Fall fishery will make a saving voyage, and that you may have a bumper next year."[26]

Although the first whale processed at Aquaforte in the previous two seasons was on 18 April, *Humber* was still in St John's taking on supplies on that date in 1906.[27] Not only was the opening of the season delayed but whales were scarce when they finally did appear (fig. 6.2).[28] *Humber* took only six by the end of May, usually the most productive period for the station.[29] Similarly, by early June "only 50 whales had been taken by all the steamers … as against 185 during the same period last year."[30] It thus became necessary to make longer and more expensive voyages. Earlier in the month, for instance, *Humber* had been "in Placentia Bay all the week and got nothing."[31] Other companies with stations on the south coast were doing just as poorly: "The past week [11–17 June] is regarded as the best week of the year for whaling, and it is [sic] utter failure shows better than anything else the collapse of the industry."[32]

With his workers seriously underemployed, Ellefsen tried to offset his losses by putting twenty men to work fishing for cod.[33] *Humber* returned to Aquaforte on 24 July, having been "out all yesterday, but did not see a fish [whale]. Up to date she has but seven fish, and very poor prospects for the balance of the season."[34] Consequently, the company's shore crew again went cod fishing.[35] Whale catches did not improve by the end of September,[36] and the

Table 6.1
Whale Catchers and Catches, Newfoundland and Labrador, 1905.

			Species			
Catcher	*Blue*	*Fin*	*Hump*	*Sei*	*Sperm*	*Total*
Avalon	7	16	18	-	-	41
Puma	26	59	12	-	-	97
St Lawrence	27	39	4	-	-	70
*Cabot**	66	17	7	2	-	92
*Lynx**	68	63	3	-	3	130+
*Mic Mac**	37	34	13	-	-	84
Hump	-	47	12	-	2	61
Fin	-	23	10	-	-	33
Baccalieu	-	20	2	-	1	23
Falcon	9	19	23	-	-	51
Humber	17	25	23	-	-	65
Port Saunders	4	17	5	-	-	26
Hawk	2	24	16	-	-	45+
Cachalot	-	57	13	-	-	70
Total	263	460	161	2	6	892/888

Source: Newfoundland, *Annual [Fisheries] Report for 1905*, 1906.
*operated from at least two stations.
+ reported totals do not coincide with reported species numbers.

company decided that it could not sustain further losses. Ellefsen tried to sell the station and catcher, but many others were in the same position and his best efforts failed.[37] Job Bros. tried to sell him the guano factory they had recently built across the bay from Aquaforte, arguing that this would consolidate his operation and help the company survive.[38] The sale did not take place, and it became increasingly evident that Aquaforte would not operate in 1907.[39]

In an attempt to revitalize the industry, the government halved the cost of station licences to $750 each.[40] At the same time, the price of whale oil, "owing to the scarcity … advanced again and now stands at £22, against £13 last year."[41] Many thought that these developments would help sustain a smaller industry, especially since some stations had closed, reducing pressure on remaining stocks.[42] However, the decline continued through to 1908 when only six vessels operated from ten stations (fig. 6.1). With competition reduced, catches per vessel did improve slightly in 1907,[43] the first time in four years, primarily because stations on the northeast coast and at Hawke Harbour, Labrador, had particularly successful seasons.[44] As a result the Cabot Steam Whaling Company and the Labrador Whaling and Manufacturing Company paid dividends of 12 per cent and 15 per cent respectively.[45]

Table 6.2
Catch and Production, Newfoundland and Labrador Shore-Stations, 1906–07

Station	*Whales*		*Oil (Gals.)*		*Guano (Tons)*		*Bone (Tons)*	
	1906	*1907*	*1906*	*1907*	*1906*	*1907*	*1906*	*1907*
Balaena	24	32	59,392	33,831	212	–	103	4
Snook's Arm	*	75	–	63,024	–	–	–	7
Chaleur Bay	58	37	60,178	36,546	188	–	113	5
Rose-Au-Rue	67	84	44,653	63,720	182	–	106	82.75
Beaverton	24	42	31,006	46,181	95	–	50	7
St Lawrence	21	30	20,520	26,422	78	93.5	35	49
Cape Broyle	3	2	2,898	2,436	9	–	7	0.25
Cape Charles	25	31	16,640	23,814	–	19	62	3
Dublin Cove	27	27	30,080	25,956	62	70	50	50
St Mary's	4	*	3,200	–	13	–	6	–
Aquaforte	18+	3	13,824	2,495	–	–	15	0.25
Safe Harbour	21	*	28,300	–	73	–	32	–
Harbour Grace	17	*	22,680	–	65	–	16	–
Hawke's Bay	40	31	83,160	36,000	146	119	69	55
Hawke Harbour	60	63	75,600	73,250	130	125	52	200
Trinity	*	24	–	19,600	–	25	–	40
Total	409	481	492,131	453,275	1,253	451.50	718	503.50

Sources: Newfoundland, *Annual [Fisheries] Report for 1906*, 1907; ibid. [1907] 1908.
*Station not operating.
+Company records show 17 whales killed.

Relatively good catches continued in 1908. By late August *Cachalot* had brought fifty-three whales into Hawke Harbour, "an excellent showing,"[46] reaching one hundred by the end of the season.[47] All other companies, except for the Hawke's Bay Whaling Company, also matched their numbers of the previous season.[48]

IMPLICATIONS OF THE 1905–08 WHALING SEASONS

With virtually no opportunity to recover their original investment by selling catchers, factories, and equipment, many companies adopted new survival strategies. The most productive stations developed partnerships so that they could use fewer vessels, and joint ventures were formed with foreign companies. Several Newfoundland and Labrador personnel and catchers, for example, played important roles in establishing shore-station whaling in British

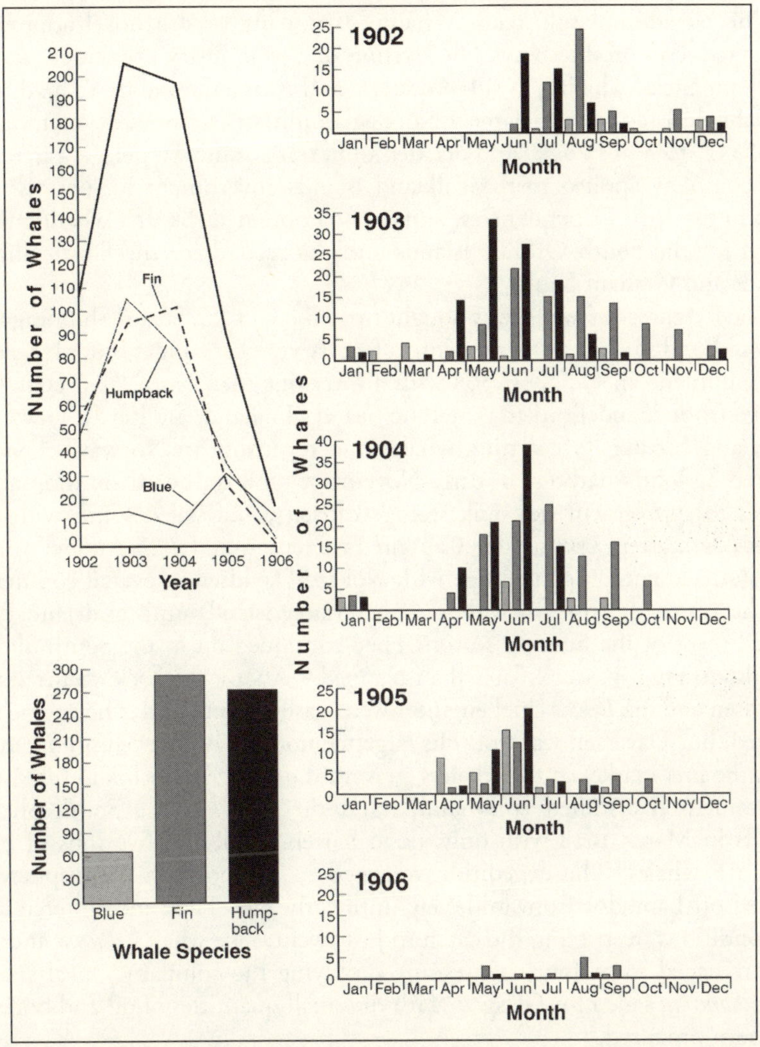

Fig. 6.2 Aquaforte whaling station: Monthly, annual, and total catches by species (1902–06). *Source*: Ellefsen Papers

Columbia, Japan, and to a lesser extent, Alaska and the west coast of the United States. Six idle Newfoundland catchers, *Humber, Falcon, Avalon, Neptune, Fin,* and *Baccalieu*, were sold to newly established Japanese whaling companies between 1904 and 1909 (appendix 4).[49] A few Newfoundlanders helped sail the vessels to the Orient, but most were crewed by experienced Norwegian masters, officers, and engineers.[50] Many of the locals who had worked on the stations and as ordinary seamen on the catchers returned to their traditional fisheries or were employed in the construction of a new pulp and paper mill at Grand Falls.[51]

The Newfoundland Steam Whaling Company tried a novel approach to offset reduced productivity. The decline of the industry coincided with the development of whaling in sub-Antarctic and Antarctic waters.[52] The directors probably learned of these largely Norwegian initiatives from crewmen on their catchers *Lynx* and *Puma*,[53] and decided to fit out a southern pelagic excursion.[54] The company applied to the Falkland Islands government for permission to hunt in the then Dependencies, and in 1908 obtained the first whaling licence issued for the South Orkney Islands and the second for the South Shetland Islands and Graham Land.[55]

The freighter *Sobraon*[56] was bought from the Gordon Steam Shipping Company of London[57] and converted into a factory vessel at Sandefjord.[58] It sailed for Newfoundland in October 1908 with a forty-one-man Norwegian crew[59] on a voyage from "Sandefjord to Antarctic Sea via England, St. John's, Newfoundland, and further … catching whales and returning to Norway."[60] *Sobraon* entered St John's harbour in early November, took on board six thousand oil barrels to supplement her bulk storage capacity, and sailed south with Einar Egenes as master. *Lynx* under Captain Emmensen and *Puma* under Captain Christofferson accompanied her, with Nokard Davidsen in overall command.[61]

The vessels did not reach the South Orkney Islands until mid-January and missed most of the hunting season. They continued on to the South Shetland Islands, arriving six weeks later than planned.[62] Another setback occurred when Davidsen and the *Lynx*'s chief engineer were washed overboard. The engineer was rescued, but Davidsen was lost. His large memorial was unveiled on 9 February 1908, the first marker in the whalers' graveyard on Deception Island (ill. 6.1).[63]

Emmensen assumed command,[64] and the fleet left the South Shetland Islands in March 1908 with only 3,090 barrels of oil and five tons of baleen from 116 whales. The expedition returned to the northern hemisphere and hunted off Labrador from mid-May until early July. They encountered severe ice conditions, restricting the catch to just twenty-five whales. *Puma* and *Lynx* then returned to St John's to resume supplying the company's local stations, while *Sobraon* sailed for Glasgow with the small quantities of oil and bone that had been procured.[65]

The deployment of a factory vessel to process whales in both hemispheres in the same season had been intended to revitalize the Newfoundland and Labrador industry, but the initiative did not meet expectations. Only 141 whales were killed for some 4,000 barrels of oil, an unprofitable production despite rising oil prices.[66] Heavy ice conditions were primarily blamed for the failure in the south, but the death of the experienced Davidsen undoubtedly affected operating efficiency. Some of the oil also deteriorated in transit and likely sold at a price lower than expected.[67] *Sobraon* made no further voyages for the Newfoundland Steam Whaling Company. Bought by Amadeus Andresen, it was renamed *Orion* and worked as a factory vessel off the west coast of South America.[68] The Newfoundland Steam Whaling Company was subsequently reorganized as the Rose-au-Rue Whaling Co. Ltd in April 1910 and ceased operations at the end of the 1915 season.[69]

Ill. 6.1 Memorial service for Captain Nokard Davidsen, Deception Island, 9 February 1908
Source: Sørensen 1912

Another initiative was more successful. With British Columbia whaling expanding rapidly, catchers, equipment, and experienced personnel were required. These became available as the Newfoundland and Labrador whaling industry declined. A significant number of skilled Newfoundlanders thus temporarily or permanently relocated to the West Coast of Canada (see chapter 8).

CONSOLIDATION AND CLOSURE, 1909–17

The reduction in catches, catchers, and stations that occurred after 1905 continued until the Newfoundland and Labrador industry closed temporarily at the end of the 1916 season (table 6.3). Some companies had been able to operate for longer periods than others and enjoyed relative stability from about 1909 (fig. 6.1). This was probably because of reduced competition, better hunting conditions, and larger catches for the more northerly stations.[70] But the improvement came too late for them to make enough money to meet expenses and recoup their investments in catchers and stations.[71] Attempts to increase productivity by using different combinations of vessels, stations, and whaling grounds from season to season generated indifferent results, and efforts to sell the remaining catchers and stations[72] were unsuccessful.[73] Premises were abandoned, and men and vessels transferred to other activities.[74]

By 1909 the industry had been reduced to six companies using seven catchers to supply nine factories (fig. 6.1). The Cabot Steam Whaling Company

Table 6.3

Newfoundland and Labrador Whaling Industry, 1909–17

Company	*Catchers*	Stations *1909*	*1910*	*1911*	*1912*	*1913*	*1914*	*1915*	*1916*	*1917*
Cabot Steam Whaling Co. Ltd (Harvey and Co. Ltd)	*Cabot*	Balaena Snook's Arm	Balaena Snook's Arm	Balaena Snook's Arm	Balaena Snook's Arm	Balaena Snook's Arm	Balaena Snook's Arm	–	–	–
Newfoundland Steam Whaling Co. Ltd	*Puma/Lynx*	Rose-au-Rue	Rose-au-Rue	Rose-au-Rue	Rose-au-Rue	Rose-au-Rue	Rose-au-Rue	Rose-au-Rue	–	–
Cape Broyle Whaling and Trading Co. Ltd (Bowring Bros. Ltd)	*Hawk*	Cape Charles	Cape Charles Cape Broyle	Cape Broyle	Cape Broyle	–	–	–	–	–
Phoenix Whaling and Manufacturing Co. Ltd	*Hump*	Dublin Cove Trinity	Dublin Cove Trinity	Dublin Cove Trinity	Dublin Cove Trinity	Dublin Cove Trinity	Dublin Cove Trinity	–	–	–
Hawke's Bay Whaling Co. Ltd	*Port Saunders*	Hawke's Bay	Hawke's Bay Harbour Grace	Hawke's Bay Harbour Grace	Harbour Grace	Harbour Grace	–	–	–	–
Labrador Whaling and Manufacturing Co. Ltd	*Cachalot I* (lost 1910) *Cachalot II*	Hawke Harbour	Hawke Harbour	Hawke Harbour	Hawke Harbour	Hawke Harbour	Hawke Harbour	Hawke Harbour	Hawke Harbour	–

Source: Sanger and Dickinson 2000.
Dashes indicate not operating.

operated *Cabot* from Balaena at the beginning of the season, and then moved it north to Snook's Arm. The company continued to use this strategy until it ceased operations in 1914.[75] After the failure of the *Sobraon* experiment, the Newfoundland Steam Whaling Company enjoyed reasonable success as the Rose-au-Rue Whaling Company to 1915 by deploying *Lynx* and *Puma* from Rose-au-Rue only.[76] The Cape Broyle Whaling Company used *Hawk* at Cape Charles, Labrador, in 1909 and reopened Cape Broyle the following year. The additional expense of operating both stations resulted in the closure of Cape Charles at the end of the 1910 season, followed by Cape Broyle in 1912.[77] The Phoenix Whaling Company used *Hump* at Dublin Cove on the south coast before having her hunt from Trinity between 1909 and 1914.[78] The Hawke's Bay Whaling Company, the least productive towards the end of this first whaling phase, reopened the Harbour Grace station in 1910 after *Port Saunders*'s lack of success on the west coast during 1909. This move also failed, and their two stations closed in 1911 and 1913 respectively.[79]

The most successful whaling company between 1909 and 1916 continued to be the Labrador Whaling and Manufacturing Company. Hunting off southern Labrador throughout this period was especially productive,[80] enabling the company to order a replacement catcher from Norway after *Cachalot* ran aground on Belle Isle in late August 1910 at the height of the season.[81] Also christened *Cachalot*, the new vessel was more powerful and ice-strengthened,[82] allowing the company to hunt where others could not.[83] By 1916, however, catches and oil prices had declined to the point that the company president, Daniel Ryan, decided to stop whaling. The Hawke Harbour station, catcher, and equipment were sold,[84] bringing the industry to a close, albeit temporarily.

In a presentation to the Newfoundland House of Assembly of which he was a member, Ryan succinctly summarized the rapid rise and fall of the industry over two decades: "Take the whole business. We had some parties introduced this, and they got along very well. Then we had two or three factories made a big success and then all wanted to get into it. I remember at the time being a member of the Government when at one time we had twenty-three applications for those factories in one batch. Fortunately for the most of these people the applications were not granted. The few who got them lost all."[85] The first exploitation-depletion cycle of the modern Newfoundland and Labrador shore-based whaling industry had lasted just nineteen years.

CHAPTER 7

Expansion and Decline at Aquaforte

Commercial shore-station whaling in Newfoundland and Labrador was not an activity of large companies that preserved their records. Most were short-lived, opportunistic endeavours seeking quick profits. A comprehensive set of records seems to exist for just one station, that at Aquaforte. The Ellefsen Papers provide the basis for a case study of the operations of early twentieth-century commercial whaling.[1]

NORWEGIAN ORIGINS

The new technologies developed by Svend Foyn and others soon produced a Norwegian-dominated shore-station whaling industry on the Finnmark coast. A potential setback to their monopoly occurred when the Deutsche Polar Schiffahrts Gesellschaft of Hamburg built a whaling station on the south side of Varanger Fjord opposite Foyn's factory at Vadsø. Foyn petitioned the king of Norway and Sweden to prohibit this foreign intervention so that he could "create a new industry for Norway."[2]

On 14 January 1873 Foyn was granted exclusive whaling rights for ten years in Norwegian waters. This not only eliminated the German threat but also discouraged domestic competition.[3] To redress declining relationships with other local investors brought on by this monopoly, Foyn awarded whaling concessions to four Norwegian companies in 1880. One of these, the Stokke Interessentskap for Hvalfangst, had been established by Andreas Ellefsen of Stokke, a community between Tønsberg and Sandefjord (fig. 7.1) on a stretch of coast that "provided recruits not only to practically all the crews of the Norwegian whaling fleet, but also a substantial proportion of the crews on board the whaling fleets of other nations; and for fifty years Sandefjord was the world centre

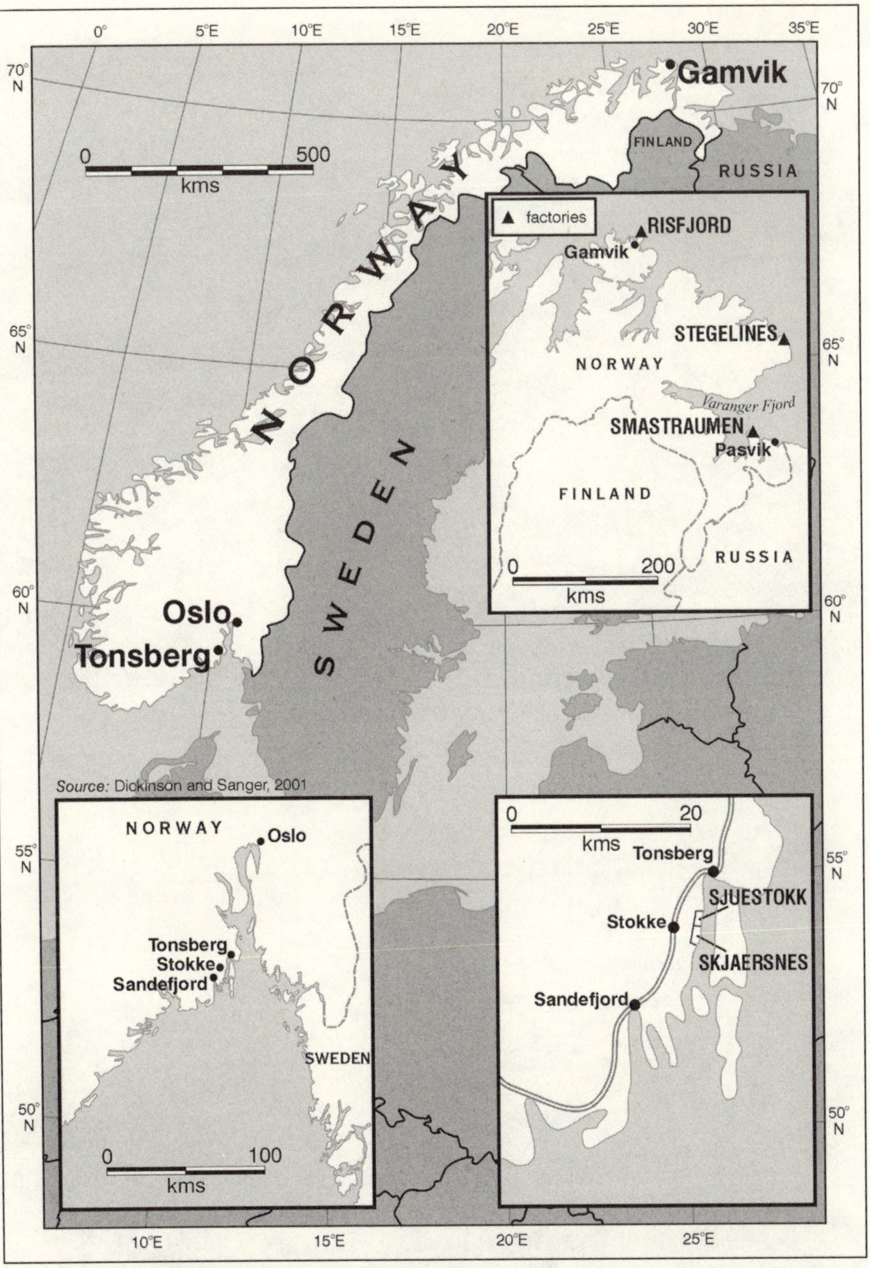

Fig. 7.1 Ellefsen family origins and whaling stations *Source*: Dickinson and Sanger 2001

for the conversion, fitting out, and equipping of whaling vessels, floating factories as well as catchers."[4]

The Ellefsens had become affluent landowners by 1814 on the two adjoining agricultural estates of Skjaersnes (ill. 7.1) and Sjuestokk (ill. 7.2), developed from the profits of various maritime ventures.[5] Andreas Ellefsen's familiarity with his neighbour Svend Foyn's whaling activities, and his position as major

Ill. 7.1 Skjaersnes, Sjuestokk in background (Ellefsen Papers)

Ill. 7.2 Sjuestokk, main house, c. 1900 (Ellefsen Papers)

shareholder in the Stokke Interessentskap for Hvalfangst, gave the family the background and incentive to enter the industry. In 1881 they built their first station at Stegelines in Iceland and later (1883–92) expanded to Småstraumen and then (1892–1901) to Gamvik.

Finnmark catches decreased rapidly from the late 1890s, and hunting was banned entirely in 1905. As the decline progressed, Norwegian whaling entrepreneurs sought other opportunities. Thirty years of Finnmark and Icelandic whaling had thus been "a nursery for the training of the crews and testing of the equipment that made it possible to develop modern catching from a local Norwegian phenomenon to a global industry."[6] One such opportunity lay in Newfoundland and Labrador, to which Andreas Ellefsen was attracted in 1901 by reports of large catches and profits.[7] He decided to enter this rapidly expanding industry by building a station at Aquaforte.

THE BEGINNING OF WHALING AT AQUAFORTE

Although the Cabot Steam Whaling Company had caught whales from Aquaforte in 1898 to be processed later at Snook's Arm, no consideration was given to building a station there. Three years later, however, the *Evening Herald* reported:

> The whale fishing industry on this coast will likely expand in the near future, and much foreign capital will be invested in it. Last week two Swedish gentlemen visited Aquaforte to negotiate for the purpose [purchase] of an extensive waterside premises there, to make a start in the enterprise as early as possible. If arrangements can be finalized for the taking over of the property alluded to, they will build very large stores and deep water piers, with a coal shed of a capacity for 1,000 tons. A new steamer of large size will be immediately begun for the prosecution of the voyage.[8]

The two "Swedish gentlemen" were Andreas Ellefsen and Ragnvald Berg. Berg was the owner of an agricultural estate adjacent to the Ellefsens at Skjaersnes and eventually became a major shareholder in the new whaling venture (table 7.1).[9] On 18 November 1901 the two bought fourteen acres of waterfront land in Aquaforte from Peter, Robert, and Arthur Carter for their Atlantic Whaling and Manufacturing Co. Ltd, including "The Island" and "Montagues Peninsula" (fig. 5.1).[10] Nylands Mekaniske Vaerksted of Christiana (Oslo) was contracted to build a new catcher, *Humber* (table 7.2).[11]

News that a Norwegian company was becoming directly involved in the growing local whaling industry did not meet with unanimous approval. A letter to the *Evening Herald* from a "Newfoundlander," for example, complained:

> The sooner these Norwegian gentleman understand that the fisheries of Newfoundland belong to the fishermen of Newfoundland and not to Norway, and that under no possible circumstance can such an enterprise be permitted here, the better for all

concerned. If the Norwegians can invade this colony and catch its whales, what hinders that they should also take over the BEST OF OUR FISHERIES? Why should they not come and starve out our hardy toilers altogether and destroy our entire marine wealth – our cod, our herring, our lobsters, etc., "by their competition."

The writer concluded "that at all hazards the foreigners must be kept off the limited patrimony which a niggardly fate still permits us to control." Such comments were representative of the public debate "among all classes of citizens over the proposal to establish a Norwegian whaling company," the general consensus being that "the company [should] have half its stock held here."[12]

Anders Ellefsen was thus obliged to respond publicly:

Sir, – Your editorial in this evening's issue [30 October 1901] is rather misleading, as happens to have been noticed by one who know [sic] something about it. Allow me therefore space to make a few observations on the subject. That "the foreign whaling concern on [sic] Iceland," alluded to, belongs to my uncle [Hans Ellefsen] … He has two factories, one was burned, but it is intended to be rebuilt; that he has decided to come to Nfld. is not true. I, undersigned, have come and with my own capital intend

Table 7.1
Major Shareholders in the Atlantic Whaling and Manufacturing Co. Ltd, 30 October 1901

Shareholder	*Occupation*	*Shares*	*Value (Kr)*
Andreas Ellefsen	farmer/whaler/shipowner	2	30,000
Hans Ellefsen	whaler	2	30,000
Fritjof Ellefsen	whaler	1	15,000
Ragnvald Berg	farmer/investor	1.8	27,5000
(and "5 Others")		6.8	205,000
Totals: 9		13.6	102,500

Source: Shareholder Agreement, 30 October 1901, Ellefsen Papers.

Table 7.2
Construction and Financing for Humber

Payment (Kr)	Activity
20,000	Contract signed
20,000	Construction started
20,000	Installation of engine
19,000	Completion date: 30 April 1902
79,000	Total cost

Sources: Contract Agreement, 16 November 1901; 13 May 1902; 7 June 1902, Ellefsen Papers.

to erect a Whaling Establishment in Nfld., and subsequently to become a British subject. From parties that I have sought information, I have been well received, and even from high authorities have expressions of "success" been tendered to me, for which I am very thankful.[13]

A few Newfoundlanders did support the Norwegian enterprise, arguing that it would provide new employment, increase financial benefits to local construction companies and suppliers, and generate licence and tax revenues for the government. The interest of the latter is shown in its response to a letter from the company's solicitor, C. O'Neill Conroy, in which he requested information on the conditions Anders Ellefsen had to satisfy before he could begin whaling at Aquaforte:

[It] is perfectly competent for your client to carry on such enterprise in this colony without obtaining a license from the Government.

In this Colony, a resident foreigner has as much right under the law to engage in business, as a British subject, and will receive at the hands of the Government equal protection in the conduct of his business.

The Government will be very pleased indeed to encourage by every legitimate means not only your client, but any other foreigners who may be desirous of enterprising business here.[14]

The way was thus clear for the construction and operation of the station, albeit with stringent controls on the employment of "any person not being a British subject."[15]

ESTABLISHMENT OF THE AQUAFORTE WHALING STATION, 1902

Construction began in March 1902 using materials brought from St John's on the schooners *Fashoda*, *Nonpareil*, and *Mary Star* and the coastal steamer *Glencoe*. The factory, coal store, oil tanks, and living quarters were projected to be completed by 25 May 1902.[16] *Humber* (ill. 4.5) arrived at St John's from Norway on 6 June 1902, began hunting 11 days later (ill. 7.3), and captured 108 whales by the end of the season in December (figs. 7.2, 7.3). Most were taken between Cape Race and Bay Bulls, and some off Cape Bonavista.[17]

Despite complaints about the stench of decaying carcasses, most Newfoundlanders welcomed the Ellefsens' presence, since the new station significantly improved local employment opportunities (ills. 7.4, 7.5), with "all the male residents of the place ... working in the factory and only four boats are engaged in the cod fishery."[18] Incomes that often exceeded $20 a month were a welcome alternative to the uncertainties of the inshore fisheries. As on other stations, local men continued to be employed in less-skilled positions onshore and earned significantly less than the Norwegians on *Humber* (appendices 5, 6).[19]

Ill. 7.3 Newly launched *Humber*, Nylands Mekaniske Verksted, Christiania (Oslo), 1902 (Ellefsen Papers)

St John's suppliers also derived considerable benefit from the company, as when *Humber* took "a large stock of meats from W. Campbell and other supplies" to the station in early September. Additionally, the government collected duties on oil and bone exports, including "the first full cargo of whale oil ever sent from this colony ... for Grangemouth, Scotland" on the Norwegian steamer *Premad* in late August.[20] The first operational season was thus an unqualified success (fig. 6.2), with *Humber* averaging "over one a day since starting, 93 whales for less than three months fishing. Last week [mid-September] was a poor one, only four whales being taken and three of these were killed near Cape Bonavista and towed south to the factory. The fourth was a sulphur-bottom [blue whale], and very valuable, killed on the banks off Aquaforte."[21] Catches declined towards the end of the year, with the 108 whales killed by the end of December accounting for 24 per cent of the 450 taken by all four vessels then hunting in a booming year for the expanding industry.[22]

THE 1903–04 SEASONS

The commitment of the Ellefsens to Newfoundland was enhanced by the announcement that Anders would marry a local girl. Eda Hopkins Melvin (ill. 7.6) was born on 23 April 1882 in St John's, the fourth of six children of Phoebe Jane Martin and James William Melvin, an accountant with Horwood and Co., a lumber company. James Melvin died on 7 July 1900, making it necessary for Phoebe to open her home at 8 Gower Street to boarders, including Anders

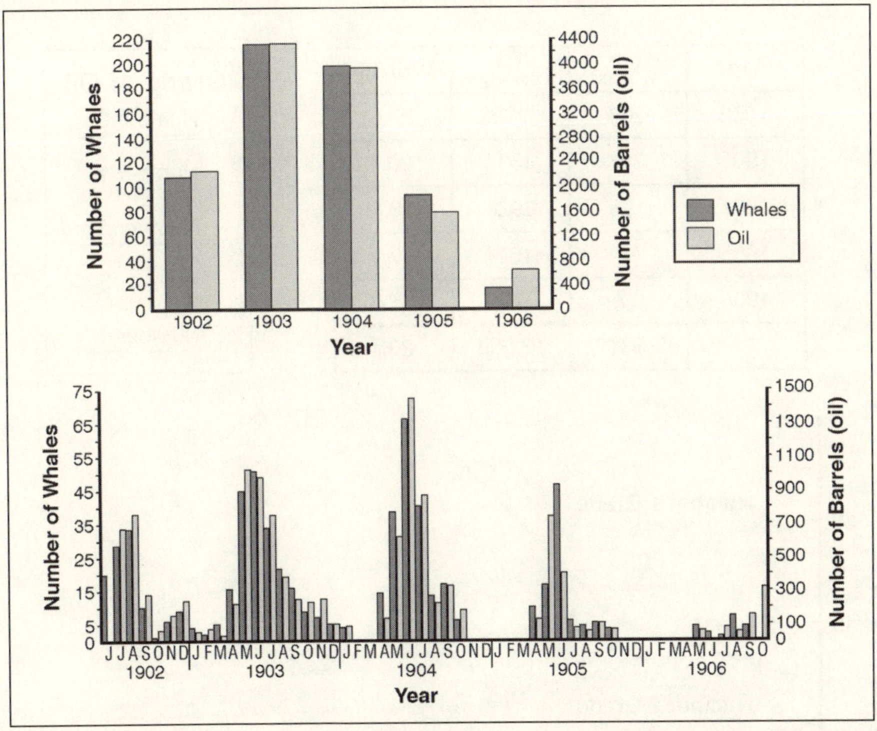

Fig. 7.2 Aquaforte whaling station: monthly and annual catches and oil production (1902–06)
Source: Ellefsen Papers

Ellefsen when he visited St John's. A relationship with Eda began when he hired her from her position as a children's governess to help him learn English.[23]

The couple were married by the Rev. H.B. Cartwright at the Cathedral of St John the Baptist in St John's on 5 January 1903.[24] Eda's brother and sister, James and Beatrice Melvin, and Anders's brother Henrik were among the witnesses.[25] The newlyweds took up residence at Aquaforte, only to suffer an immediate setback. A fire broke out on 8 January and their "dwelling house was completely consumed. The fire occurred through the explosion of a lamp and the occupants had barely time to save their lives. They lost everything only what they had on. Fortunately the east wind and wet weather saved the whale factory. Crowds of people gathered, and with buckets did all they could to save the house, but failed."[26] Anders and Eda Ellefsen returned to St John's on *Humber* to arrange for their house to be rebuilt (ill. 7.7).[27]

Humber, meanwhile, continued hunting successfully,[28] only stopping for a short time in mid-September when a lack of whales provided the opportunity to send the vessel to St John's for an annual boiler inspection. As in 1902, *Humber* then resumed the hunt "in the northern bays," and one month later had

Year	Number of Whales	Oil (Barrels)	Average
1902	108	2223	20.6
1903	216	4333	20.1
1904	198	3928	19.8
1905	92	1571	17.1
1906	17	603	35.5
	631	12,658	20.1

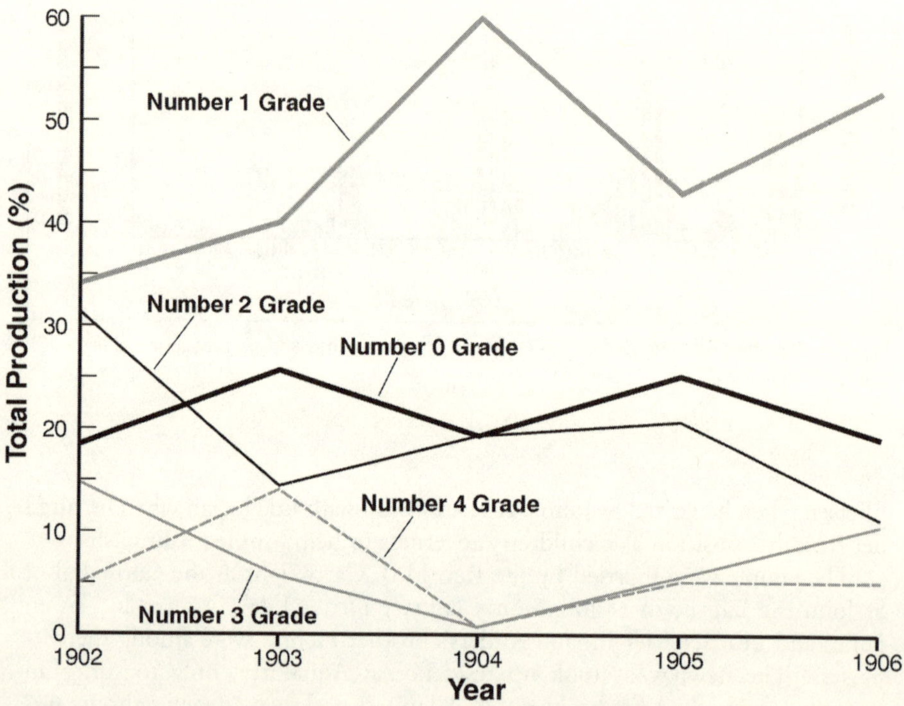

Fig. 7.3 Aquaforte whaling station annual oil production (1902–06)
Source: Ellefsen Papers

Ill. 7.4 Crew of *Humber* and station workers, Aquaforte, c. 1903. Front (fourth from right), James C. Melvin, brother of Eda; rear (second from left), Henrik Ellefsen; rear (right), Anders Ellefsen (Ellefsen Papers)

Ill. 7.5 Aquaforte whaling station, c. 1904 (Ellefsen Papers)

killed two hundred. Whales then became scarce, and poor weather made hunting difficult. *Humber* returned to St John's in October "for repairs and general clean up and painting."[29] At least 216 whales were processed at Aquaforte in 1903 (fig. 6.2), fulfilling the promise of the previous season.[30]

The company still faced one serious problem. The 1902 whaling act specified that the government could "from time to time" enact regulations to dispose of whale carcasses and prevent them becoming "a nuisance, or in any manner polluting the waters so as to be injurious to the public health, or to the fisheries of this Colony."[31] Carcasses, however, continued to be towed to the guano plant recently built at the nearby Cape Broyle station, perhaps because Anders Ellefsen did not wish to incur the expense of building his own. In May 1903, for example, he had asked Dr Rissmüller if he "would make it possible to pull four carcasses during this year from Aquaforte to Cape Broyle, so that I can tell the Government if it should enquire, that arrangements are being made to dispose of the carcasses."[32] Heavy catching between April and August kept the Aquaforte station and *Humber* too busy to comply with this arrangement, and carcasses continued to be disposed of at sea.[33]

Primarily in response to the situation at Aquaforte,[34] the government sent letters to all station owners in late 1903 informing them that "the Government have decided ... [to] make it compulsory on the part of all holders of whaling licences to manufacture into a commercial commodity the whole of the carcass of the whale. In other words to prevent [whaling] without having a station or factory for the conversion of the offal and carcass of the whale into oil and guano."[35] Ellefsen's precipitating role in this decision is shown in a letter he received from his St John's lawyer: "I have had information today from a private source, that the Government is contemplating closing down your factory on the ground that you have no guano factory in connection therewith. I told my informant that you had arranged with Dr. Rissmuller for the erection of a suitable plant as soon as it was possible to obtain the necessary machinery, and I hope that this will be reported in the right quarter, where it will do good."[36] It was thus imperative that whole carcass utilization be introduced to Aquaforte before the opening of the 1904 whaling season.

Rissmüller's agreement to buy carcasses from Ellefsen had also called for him to build a guano plant at Aquaforte to begin operations before 1 March 1904.[37] Ellefsen, however, tried to find other, presumably more lucrative, markets for the carcasses, and so did not sign the contract until mid-October 1903.[38] In particular, he negotiated with T.R. Hunter, manager of the Huffman Manufacturing Company of Philadelphia, a major purchaser of baleen. Hunter had arrived in St John's soon after Ellefsen finalized arrangements with Rissmüller and proposed instead to take "all of the jaw bones [baleen] ... at fifty dollars per ton."[39] Despite his agreement with Rissmüller, Ellefsen found this offer too tempting to refuse. Not surprisingly, Rissmüller strongly objected and claimed that Ellefsen had also approached Bowring Brothers, his partners in the whaling station at Cape Broyle, "behind my back and acted in every other way in a manner no decent businessman would employ."[40]

Ill. 7.6 Eda Hopkins Melvin, c. 1900 (B. Melvin)

Ill. 7.7 Manager's residence and Aquaforte whaling station, c. 1904 (Ellefsen Papers)

Ill. 7.8 Dr L. Rissmüller's guano factory, south side of Aquaforte harbour, opposite Ellefsens' whaling station, c. 1904 (Ellefsen Papers)

Hunter became aware of Rissmüller's protests and reassured Ellefsen that "the Dr. [Rissmüller] said he would stop you from selling me, but he can't do that until he has built his plant at your place … If you should have any trouble with the Dr. we will put up a plant at your place and use up your waste."[41] Ellefsen informed Rissmüller that their agreement was no longer valid and that he intended to sell his carcasses to Huffman. Rissmüller responded immediately with great indignation:

I was very much surprised that you broke your agreement. I thought you might have forgotten about it, but would remember it when I reminded you about the details. I expected you would apologise, and try to make your mistake good, if it is your power at all. Instead you dare to write me a most malicious letter full of the meanest and hideous insinuations. When you wrote it, you must not have known what you are doing … Did it ever come to your mind, that there would not be a decent price in the States for the bones in question if I would ship my output there? The demand at a high price only existed, by me keeping the masses of bones out of the market in the States. I *did not* ship a single pound there or anywhere else … I have created the demand for these bones by a large output of cash of my own and by sacrificing a good deal of my time.[42]

Ellefsen refused to back down, however, and began shipping bone to Philadelphia in early July 1903.[43] Unfortunately, Huffman in turn refused to honour the terms originally negotiated, informing Ellefsen on 23 October 1903 that "Mr. Hunter was [only] authorized to purchase 25 tons of jaws and we

[will] carry out any contract he might make to that extent."[44] Ellefsen had no choice but to accept Rissmüller's earlier offer to build a guano plant at Aquaforte.[45] The government was thus satisfied that the new plant would "protect the said Ellefsen from any penalty under Section 9 of the Whaling Industry Act, 1902."[46] By mid-March, "Job Bros. & Co., managers of the Atlantic Mfg. Co., ... [were] erecting a new fertilizing factory in Aquaforte in connection with the whaling factory there. It was begun last week and 60 men are working on it and will hurry it to completion quickly."[47] The new facility was a substantial waterfront complex of buildings, wharfs, and slipways on the south side of Aquaforte harbour opposite the whaling station (ill. 7.8).

The company resumed whaling in full compliance with government regulations, and its catch of 198 whales, the most of any of the ten stations operating that year, contributed significantly to making 1904 the peak season for the industry. Anders's brothers Henrik, Arne, Georg, and Peder now joined him. Their collective expertise as station owners and operators, catcher captains, gunners, and inspectors heavily influenced the development of Newfoundland and Labrador shore-station whaling.

DECLINE, 1905–08

Preparations for the 1905 season began in late March, soon after Anders and Henrik Ellefsen and their families returned from a visit to Norway (ill. 7.9).[48] Anders was immediately forced to request additional operating capital from home due to a decline in oil and bone prices. He needed Kr 10,000 "by June 15th, as there are 3-month bills of acceptance that fall due on that date, so be sure that the money is sent early enough to arrive here on time."[49] Ragnvald Berg was asked to convince the shareholders of this necessity, since "it's terrible when your pockets are empty."[50]

Humber took supplies and equipment to the station and caught the first whale, "a fine sulphur bottom," on 18 April.[51] Although plentiful in the first half of June, and "larger than any taken last year," only forty-six were caught by early June, "about half the total they had at this time last year." In fact whales became so scarce that only one was caught between 18 June and 10 July, when it was reported that whaling was "slack south." Henrik Ellefsen took *Humber* "north to the Funks ... in quest of cetaceans, having covered the coast from Cape Pine [south of St John's] to Cape Bonavista, for a radius of 80 miles off shore and not seen a spout."[52] There was little improvement in August,[53] and it was not until late fall that whales reappeared off Aquaforte. Even then *Humber* had to extend her hunting range into Conception Bay and off Cape Bonavista. These longer voyages increased costs and reduced profits. Only sixty-five whales were processed at Aquaforte during 1905, barely 30 per cent of the 1904 catch.[54] Stocks clearly could not support the high catches of the previous seasons and the increased hunting pressure caused by the unregulated construction of new stations.

Not only was the opening of the 1906 season delayed but whales were few when they finally did appear. By 24 July *Humber* had "but seven fish, and very poor prospects for the balance of the season." Just four were taken in August.[55] There was no improvement in September, and the Ellefsens and their Norwegian partners decided that they could not accept further losses. As early as mid-August, for example, Berg informed Anders that they must seriously consider withdrawing from Newfoundland:

> We must now soon have a board meeting … and take these matters under consideration … This is a sorry state of affairs, but there is no other option – we must quit – it is costing money every day and no catch … After the meeting, you will be given clear instructions as to what should be done … I have had enough of this, so I am not able to take part in anything that requires money, I am sorry to say. I think we must just stay calm and save as much as possible of the pieces.[56]

Plans were made to sell or charter *Humber*, petition the government to eliminate annual licences, and inquire if these fees would be required if stocks later revived. The company also considered selling the station: "You had best prepare to sell all assets that are saleable – the buildings should probably also be sold. In short, everything that is not absolutely necessary for the re-opening of the firm, if so should happen."[57]

Despite the pace at which the industry was declining, the government was unsympathetic. Anders Ellefsen was told "you can close your factory if you wish, but while you hold a licence you have to pay."[58] Although they recognized that it would be difficult to find a buyer for the station, the Ellefsens made the decision to stop whaling. Andreas informed his son on 2 November 1906 that "we must sell factory as quickly as possible,"[59] and an advertisement was placed in local newspapers:

> *Whaling Property for Sale*
>
> Tenders will be received for the undersigned … for the purchase of the Land, with Whaling Factory, Dwelling House, Bunk House, Boiler House, Coal Shed, Store, Stable, Forge, Slip, Steamer Slip, Machinery and Plant, and Tools and Utensils of the undersigned at Aquaforte. The land contains about 14 acres.
>
> Anders Ellefsen
> Crosbie Hotel [St John's][60]

In the absence of acceptable offers, the Ellefsen family decided to make a bid of their own at an auction on 5 December.[61] Auctioneer Thomas Clift was instructed "not to give any information to newspapers or others concerning same,"[62] and the Ellefsens reluctantly retained ownership of the station, 721 oil casks, and 15 tons of coal for $4,113.80.[63]

Elsewhere, however, whaling was expanding and the family was able to sell *Humber* to a new company in Japan.[64] The sale followed an enquiry by Robert

Ill. 7.9 Ellefsen family gathering, Sjuestokk (1904–05). Front, Anders's Newfoundland-born daughter, Marie; rear (left to right), Anders's brother, Henrik; wife, Eda; brother Fritjof; father, Andeas; Henrik's Newfoundland-born wife, Hett; Aunt Kristine (Ellefsen Papers)

Shankland and Co., of London, who had "direct instructions from our special agents in Japan for the purchase of several whalers. We shall be glad to hear whether you would be disposed to sell your *Humber* … This is special and serious business, and we shall be glad if you will furnish us with full particulars promptly."[65] The Norwegian broker Johan Bryde also used Anders Ellefsen as his agent to buy additional catchers for the rapidly expanding Japanese industry: "Have bought *Avalon* from Prowse. Kindly help me get crew everything of Harpoon etc. onboard. Engines ready … Wire also if *Mic Mac* or other whaleable [sic] saleable."[66] Taking advantage of the decline in the local industry, Anders purchased catchers from companies about to cease operations. Henrik and Arne Ellefsen sailed two of them to Japan[67] and briefly became involved in the development of this new industry.[68]

Having refinanced the station and sold *Humber*, the Ellefsens brought in a replacement catcher, *Othar*,[69] from Norway in 1907 (ill. 7.10). Built in Christiania (Oslo) in 1885, *Othar* was fourteen years older than *Cabot*, the first steam-powered catcher used in Newfoundland and Labrador. She was so old, in fact, that the Japanese had not considered her acceptable.[70] Nonetheless, the purchase was hailed as an act of "faith" by the Ellefsens "in the future of our whaling industry."[71]

Ill. 7.10 *Othar* on slipway, Aquaforte, c. 1907–08 (Ellefsen Papers)

Unfortunately, *Othar* did not arrive in St John's until 17 August, well past the peak catching period. Whales were then "very scarce on the Southern Shore," and *Othar* caught only two by the end of the month. That the season was not "very successful so far" was an understatement, since only three whales were processed at Aquaforte in 1907. *Othar* was laid up at Aquaforte in mid-September.[72]

The speculative and desperate attempt by the Ellefsens to continue whaling had been a dismal failure,[73] and the family decided to end its Newfoundland involvement. The station was again advertised for sale,[74] and domestic and foreign whaling companies were contacted to see if there was any interest. Prospective buyers in the United States, South Africa, and Canada were optimistically informed that whales "should come back."[75] Job Bros. was also approached to buy the station because of the "guano factory there"[76] but refused. Only one offer was received, from Messrs Brister (Bissett?) and Son of Halifax for $600, but this was rejected.[77]

Inventories of the contents of the manager's house and out-buildings, and details of machinery and supplies (appendix 7), were circulated to prospective buyers, and the vessel and property were again re-advertised.[78] All of the Ellefsens' best efforts to sell failed. With winter coming on, and increasingly anxious to return home, Anders Ellefsen somewhat despairingly telegraphed his father on 10 October 1908: "Cannot obtain an offer. May I leave?"[79] Consent was given,[80] and Anders, Eda, daughter Marie, and little Edgar (ill. 4.3), born at Aquaforte on 25 February 1906,[81] sailed for Norway on 13 November.[82] The station was left under the care of Johan Andreassen, a Norwegian, who was paid $10.75 monthly by the Ellefsens.[83]

Ill. 7.11 Friends and family celebrate the ninetieth birthday of Marie (Ellefsen) Kiserud (centre), Oslo, 11 January 1994 (B. and E. Kiserud)

The station was eventually sold in 1912 to A/S Kosmos, then operating three large combined whaling-sealing vessels in the western North Atlantic. The company owner, Christian Christensen of Sandefjord, a "leading personality in this significant phase of modern whaling,"[84] paid $US8,000,[85] finally severing all connections between the Ellefsens and Newfoundland.[86] The rise and fall of their Aquaforte operation mirrored the events that affected all participants in the initial phase of local shore-station whaling.

The Ellefsen family's involvement in whaling did not end after they closed Aquaforte. Henrik worked in Japan and South Georgia, Arne in Japan, and Hans and his son Carl at Saldanha Bay, South Africa. Anders returned to Skjaersnes, and in 1909 opened Ellefsen's Tenskruefabrikk to make harpoon time-delay fuses following Norwegian patents granted to him in 1919 and 1931. He also farmed and became mayor of Stokke between 1932 and 1941, and again in 1945 before the Nazis replaced him with a local sympathizer. A German plan to manufacture mines at their factory failed when Edgar Ellefsen, born at Aquaforte, helped the Norwegian resistance remove a critical piece of machinery. Anders died on 14 December 1953, followed by Eda (Melvin) on 23 January 1956. Eda never returned to her native Newfoundland after leaving Aquaforte in 1908.[87] Marie (Kiserud), eldest of the three children of the marriage, lived in Bygdøy, Oslo, until her death in 1995 (ill. 7.11).

CHAPTER 8

Newfoundland and British Columbia Whaling Connections

ORIGINS

Whaling probably began in British Columbia with the use of stranded animals by indigenous people, and, as in Newfoundland and Labrador, evolved during the early part of the twentieth century into a commercial shore-station industry.[1] By the time American and European traders reached the Pacific Northwest in the late eighteenth century, coastal Indians were using boats, harpoons, and sealskin floats to capture whales.[2] Commercial hunting, however, did not begin until American, English, and French whalers arrived in the mid-1830s.[3] This "traditional whaling" seriously reduced bowhead, right, and sperm whale stocks in the region. As on the Atlantic coast, the introduction of Norwegian technology enabled British Columbian whalers to begin capturing other species from 1904, just as "modern" whaling was reaching its peak in Newfoundland and Labrador.[4]

The modern industry began in British Columbia when Captain Sprott Balcom[5] (ill. 8.1) of Halifax, who had been heavily involved in south Atlantic[6] and Pacific sealing, convinced William Grant of the Victoria Sealing Company to support a whaling venture similar to that on the east coast.[7] The Pacific Whaling Company began operations from Sechart (fig. 8.1; ill. 8.2) in September 1905 with the new catcher *Orion*, brought from Norway by Sprott Balcom's brother Rueben.[8] The original investors "were amply repaid. At the first general meeting in January 1907, 23 percent dividends were announced on preferred stock and 16 percent on common shares, a very fine return on a net profit of $45,552."[9] That year the company opened stations at Page's Lagoon (Piper's Lagoon) north of Nanaimo, and at Cachalot, Kyoquot Sound. Page's Lagoon only operated for one season, but Kyuquot enjoyed early success. The compa-

Ill. 8.1 Sprott Balcom, Halifax, c. 1885 (G. Balcom)

ny's station at Rose Harbour, the first on the Queen Charlotte Islands, began processing whales in 1910.[10]

Balcom and his associates were now ready to embark on a major expansion funded "from a most solvent source: the builders of the Canadian Northern Railway."[11] Under Sprott Balcom's management, the new company, Canadian North Pacific Fisheries Ltd, obtained ten whaling licences, bought five new catchers, and built a second factory at Naden Harbour on the Queen Charlotte Islands. The company also expanded into the United States by opening a station at Bay City in Washington State.

Increased hunting pressure on local whale stocks and ineffective regulations inevitably produced catch declines similar to those that occurred a decade earlier in Newfoundland and Labrador. In 1912 the total catch fell to 1,374 from 1,806 the previous season, and 1913 "was even more disastrous; only 916 whales were landed, about half the 1911 total. In just two years the average catch per station fell from 361 whales to 183; in 1914 it declined further, to 151. Like every whaling operation, improved efficiency actually worked against the company's

Ill. 8.2 Sechart whaling station (J. Goddard)

Fig. 8.1 North-West Coast whaling stations

best interests."[12] The Canadian North Pacific Fisheries went into receivership in 1914, and from 1915 to 1918 the industry was controlled by William Schupp, the German-American[13] owner of the Victoria Whaling Company. The Canadian component of the company was reorganized as the Consolidated Whaling Corporation Ltd, and although nominally separate entities, the American and British Columbian branches effectively functioned as "a single international organization"[14] until World War II. Economic recession led to Sechart being closed in 1918, and Kyuquot along with *Orion* and *St Lawrence* were sold in 1925–26. West Coast commercial whaling continued sporadically throughout the 1930s and 1940s.[15]

The industry received a boost in 1948 when the Western Whaling Co. Ltd was formed by British Columbia (B.C.) Packers, Nelson Brothers Fisheries, and Gibson Brothers. The original plan was to reopen the Rose Harbour station, but it was decided instead to convert the Royal Canadian Air Force seaplane base at Coal Harbour. Catchers were bought and Norwegians experienced in Antarctic whaling were hired. The station was only a marginal operation, however, and in 1959 B.C. Packers "declared the whaling finished."[16]

Nevertheless, two years later the company entered into a new partnership with Taiyo Gyogyo K.K. (Taiyo Fisheries Ltd), Japan's largest whaling operation, to process whale meat for the Japanese market.[17] Reduced stocks could not support the new endeavour, and Coal Harbour closed permanently in 1967,[18] bringing more than half a century of shore-station whaling in British Columbia to a close.

INFLUENCES FROM NEWFOUNDLAND AND LABRADOR

The development of West Coast whaling was heavily influenced by the growing Newfoundland and Labrador industry.[19] Captain Sprott Balcom sailed his sealing schooner around Cape Horn to British Columbia in 1892 and later brought his wife and family overland by rail to join him in Victoria.[20] He was associated with the Victoria Sealing Company for the rest of the decade but kept his East Coast connections through his son Harry, also a sealing master.[21] Balcom was therefore likely aware of the early successes of the Cabot Steam Whaling Company, widely reported in regional newspapers. In August 1898, for example, the *Halifax Chronicle* commented that "the recent establishment of a whale oil manufacture in Notre Dame Bay calls attention to the fact whaling is now a paying venture off the coast of Newfoundland. Recently seven whales worth nearly $1,000 each were captured by a company of Norwegian whalers. There is no apparent reason why Halifax should not become an outfitting port for the whaling industry."[22]

Balcom returned to Halifax for a visit in 1900, familiarized himself with the expanding Newfoundland whaling industry,[23] and convinced the Victoria Sealing Company that their methods, equipment, and personnel could be used to develop a similar enterprise in British Columbia. The company's application to

Ill. 8.3 Dr Ludwig Rissmüller, Victoria, c. 1906 (G. Balcom)

the Canadian government for sole whaling rights on the West Coast noted that Sprott Balcom had visited Newfoundland "to look thoroughly into the question of the prosecution of the industry."[24] He spent time at Ludwig Rissmüller's (ill. 8.3) Little St Lawrence station which had started operating the previous year.[25]

Before returning to Victoria, Balcom hired Newfoundlanders experienced in building and operating whaling stations. A letter from his lawyer to the Canadian Ministry of Marine and Fisheries reported that he had "engaged the services of Mr. Charles Smith of St John's, Newfoundland, an expert in the construction and equipment of whaling factories, who at present is enroute with Capt. Sprott Balcom to the Pacific coast to supervise the construction, equipment and establishment of the necessary plant for the proper conduct of the business."[26] St John's newspapers also confirmed that "Chas. Smith … and a large number of others"[27] had left for British Columbia to construct and "manage a whaling factory – the first to be erected there,"[28] and that "a gentleman [Sprott Balcom] recently arrived here from British Columbia to obtain expert management for a whale factory about to be erected on the coast of the Province. A man [Charles Smith] well versed in the prosecution and conducting of the fishery has been secured at a salary of $1600 per annum."[29] Balcom's recruitment of local whaling personnel was also partially responsible for an unsuccessful attempt by some Newfoundlanders to develop their own stations on the West Coast and profit from this growing industry.[30]

Transfer of Regulations

Regulations introduced into the Canadian Fisheries Act to control the new industry were based on the 1902 Newfoundland act. By 1903, Newfoundlanders and others[31] were applying to the Ministry of Marine and Fisheries for approval to open whaling stations in the Gulf of St Lawrence.[32] This prompted the Canadian government to amend the legislation in August 1904 to avoid the problems then affecting the Newfoundland industry. As in Newfoundland, the Canadian regulations were intended to control station construction, hunting, and aquatic pollution and to protect local fisheries.[33] The article most responsible for developing further linkages between the East and West Coast industries, however, was the requirement that whales caught in British Columbia had to "be completely and thoroughly processed within twenty-four hours, without dumping 'noxious or deleterious matter' into the waterways."[34]

Transfer of Techniques

Techniques refined in Newfoundland by Dr Rissmüller, "the whale king of Nfld. and B.C.,"[35] made whole-carcass utilization possible. Commonly referred to as "a German-American chemist and engineer,"[36] Rissmüller had close family ties with the F.A. Rissmüller Chemische Dunger-Fabrik of Germany and maintained contact with at least one brother who lived there.[37] Marrying into a wealthy New York family,[38] he travelled extensively[39] before arriving in Newfoundland and opening guano production facilities at the Cabot Steam Whaling Company stations at Snook's Arm and Balaena. The first reference to his presence in Newfoundland is a newspaper report in September 1900 noting that "Dr. Rismuller left by train yesterday for Snook's Arm with a view of establishing a guano factory in connection with the whaling business."[40]

Rissmüller had previous experience in converting animal waste, having developed "methods for making 'guano' from dried and ground fish offal; this was sold as chicken feed and cattle fodder."[41] It is unclear what attracted him to Newfoundland, but once there he was "induced to visit some of our whale factories"[42] by John Harvey, principal investor in the Cabot Steam Whaling Company. Rissmüller quickly recognized the potential for financial rewards if he could adapt his guano production techniques and knowledge to processing the discarded whale carcasses as required by law. He visited the three stations operating in 1901[43] and by the following year had successfully modified his system. His impact on the expanding whaling industry, nearby outport communities, and the colony's economy is evident from a 1902 newspaper report:

> A gentleman who arrived here some time ago is now on the west coast [Reuben's Cove] where he has employed several hundred men for the purpose of extracting oil from the refuse carcase [sic] of whales, when the factory people have got all they want, and for the manufacture of guano. This gentleman, Dr. Rismuller [sic], has patented several methods for work of this kind and believes there is a great field in the colony for such. He proposes erecting a similar plant to the three now in operation in Fortune Bay at Snook's Arm and possibly elsewhere. Those who went from St John's with him to assist in the work of building speak in high terms of his uniform kindness to all his employees.[44]

By the end of 1902, the Newfoundland whaling stations were using Rissmüller's methods.[45] Cape Broyle, then under construction,[46] was specifically designed to have "a large factory near the site of the whale-house for the conversion of the offal and carcasses of the fish [whales] into fertilizer."[47] Rissmüller's lawyer was thus able to inform the Canadian Ministry of Marine and Fisheries on 15 November 1902 that "Dr. Rissmuller's company is doing all the scavenging work for all the whale factories."[48] His success led him to develop plans for his own fully integrated whaling and guano production station at Little St Lawrence, which was "in full swing" by early November 1903.[49] The naturalist J.G. Millais provided a further assessment of Rissmüller's impact on whaling following a visit in 1904. He considered it "an up-to-date whale factory under the immediate supervision of Dr. Rismuller … who has done more for whaling and the use of the whale products than any other living man. To him is owed the utilisation of every part of the whale, including the flesh, the blood and liver, and parts of the skin which were only regarded as wastage a few years ago."[50]

Rissmüller was thus already heavily involved in Newfoundland and Labrador whaling when Sprott Balcom arrived at his station in July 1904. By this date he had patented his Newfoundland developed and tested technologies[51] (ill. 8.4; appendix 8), which could therefore not be used in British Columbia without his involvement or permission. Neither was long in coming. The post-1904 decline of the Newfoundland industry convinced Rissmüller to look for more profitable opportunities elsewhere, and he gave his support to a consortium of Balcom's competitors who were attempting to establish their own whaling operation in British Columbia. In January 1905 he swore before a notary public in Ottawa that they had permission "for the use of my patented Whale Reduction Process in the factories to be erected … by them … [and that the method] … has been used for over three years in Newfoundland and has proved efficient."[52]

Balcom's Pacific Whaling Company, meanwhile, began operations from Sechart in September 1905 but without Rissmüller's guano production equipment. However, since the 1904 Canadian legislation required that "whale carcasses be processed and disposed of within twenty-four hours,"[53] it is not surprising that in December 1905 Rissmüller moved from Newfoundland to Sechart to become a major shareholder in the Pacific Whaling Company.[54] The formal arrangement between Rissmüller and Balcom was confirmed in a memorandum of agreement signed on 29 November 1905, wherein:

> Rissmuller transferred to the Pacific Whaling Company, in perpetuity, all rights to his processes, methods, and machinery as well as his personal services as consulting chemist and engineer. Rissmuller also agreed to train and supervise assistants … to perform the necessary "alterations" of the "whaling establishments and oil refineries of the … Company". In exchange, the company gave Rissmuller all 1,982 issued shares of its common stock. The value of the stock was $97,800, and the arrangement made Rissmuller a major shareholder.[55]

The Pacific Whaling Company now satisfied Canadian regulations, giving Balcom the opportunity to successfully argue that because his company was the only one on the coast authorized to use Rissmüller's process, it should be given a whaling monopoly. It is only fair, he claimed, "that other companies should not be allowed to use inferior processes, by which the waters would be polluted and by which the reputation of the whaling industry would be hurt."[56]

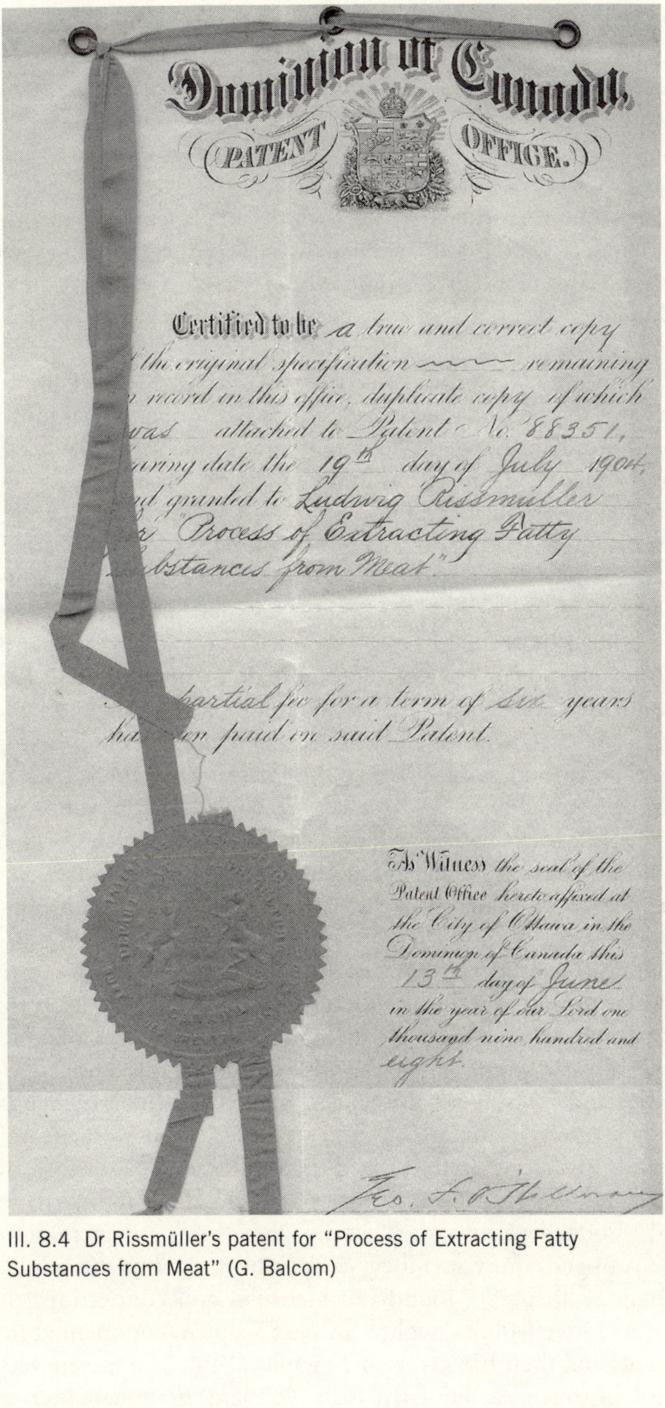
Dominion of Canada

Patent Office.

Certified to be a true and correct copy f the original specification remaining n record in this office, duplicate copy of which was attached to Patent No. 88351, aring date the 19th day of July 1904, nd granted to Ludwig Rissmüller r "Process of Extracting Fatty bstances from Meat".

partial fee for a term of Six years ha en paid on said Patent.

As Witness the seal of the Patent Office hereto affixed at the City of Ottawa in the Dominion of Canada this 13th day of June in the year of our Lord one thousand nine hundred and eight.

Ill. 8.4 Dr Rissmüller's patent for "Process of Extracting Fatty Substances from Meat" (G. Balcom)

Despite the reorganization of the Pacific Whaling Company in 1910, Rissmüller continued in his management position[57] and became a driving force in Balcom's decision to expand operations into the United States.[58] He remained active in West Coast whaling until shortly before his death in a San Diego sanitarium on 16 April 1916.[59]

Transfer of Expertise

The British Columbia industry also required management personnel. Skilled Newfoundlanders were hired to fill many supervisory positions, whereas the station labourers were usually "fifty or more Japanese and about the same number of Chinese ... secured at cheap rates through contractors, and ... typically supporting families in the Orient."[60]

The demand for Newfoundland personnel increased as Rissmüller's guano machinery was installed at the Sechart, Page's Lagoon, and Kyuquot stations in 1907.[61] Newfoundlanders were the only people available to fill the senior positions, there being "few Americans and Canadians, mostly men who have had experience in the Atlantic whaling ... [who] can be trusted with the more important positions."[62] Rissmüller's agreement with the Pacific Whaling Company, in fact, required that "his men were to operate the plants until such time as local men could be trained."[63] Besides Charles Smith and Edward "Ned" Scaplen, who had built the Little St Lawrence factory, other migrants included Alfred Gosney, Jr, previously Rissmüller's guano plant engineer at Rose-au-Rue and subsequently manager of Sechart in 1910;[64] M.F. Carrol, who supervised the sale of Little St Lawrence and later managed the Pacific Whaling Company;[65] and Captain George Le Marquand, manager of Little St Lawrence before moving to British Columbia in January 1911 "to take charge of one of the whaling company's steamers with which Dr. Rismuller is connected." Le Marquand later ran Rose Harbour, Naden Harbour, and Bay City before becoming vice-president and general manager of the Consolidated Whaling Corporation from 1920 to 1932.[66]

The commitment to the B.C. industry meant that many Newfoundland families were uprooted and virtually all links with the colony severed. Little wonder that one manager's granddaughter would later write that "for most of the women it was a wrenching experience leaving tightly-knit Newfoundland communities and crossing a continent to isolated west coast whaling stations. Though they rarely lived on the same stations, these women and children formed bonds of friendship that lasted their lifetimes."[67] Although the company's "tendership *Gray* was always carrying them back and forth on visits"[68] between the isolated stations, they usually made their permanent homes in Victoria. The younger family members easily integrated, but the original generation maintained their Newfoundland identities and connections throughout their lives. As Olive Miles, daughter of Ned Scaplen, commented in reference to her parents and their friends from Newfoundland, "They were very clannish – they kept in touch."[69] Similarly, Joan Goddard, granddaughter of William

Rolls, another of the original Newfoundland managers, recalled her mother saying "I think that was the strange part about them – they were a very close-knit group and stayed that way all their lives. They came out together from Newfoundland, worked together in the whaling, lived near each other in Victoria, and looked after each other. They were a big happy family, and only a few special outsiders ever really joined their group!"[70]

Skilled Newfoundlanders also filled middle management positions. As early as the fall of 1905, for example, five men from Cape Broyle were "making $60 a month" at Sechart.[71] Michael Cashin, a major shareholder in the company and member of the House of Assembly (MHA) for Ferryland, arranged for another eight workers to go to British Columbia in late July 1906. They were to receive "$50 per month and their passages paid to and from."[72] When Cashin's son Peter arrived in Victoria in 1911 after working his way across Canada, the "brilliant man [Rissmüller] ... was most courteous and kind" to him[73] and gave him the "job as bookkeeper on the largest whale factory they had, located at Naden Harbour on Queen Charlotte Islands."[74] In addition to Cashin and S.C. Ruck, previously manager of the Rose-au-Rue station in Newfoundland, there were "about forty men working on the main part of the plant who had been brought out there from Newfoundland. The head oil boiler was a Newfoundlander, the chief flenser was a Newfoundlander and the general foreman was also a Newfoundlander. Several of these men came from the district of Ferryland, others from St. John's, and I had known many of them before I ever thought of coming to British Columbia."[75] There were also "thirty or forty Japanese" and a similar number of Chinese, with each group having its own "separate living and eating quarters."[76]

The growing British Columbia industry provided important employment alternatives for a large number of skilled Newfoundlanders affected by the decline of their own industry. In 1907, for example, the Pacific Whaling Company was prepared to pay "good wages" and "guarantee twelve months work" to a cooper of "sober habits."[77] That same year, four men left for British Columbia, reportedly with "several years experience at that business [whaling] and without a doubt these are the men that are wanted out there."[78] In 1908 George Reid and F. Elliott returned to Newfoundland from British Columbia, reporting Dr Rissmüller to be "up to his eyes in business ... and work plentiful and times good."[79]

The rapid decline of the East Coast industry and the lack of alternative employment there meant that by World War I "whalemen" from Newfoundland could be hired as cheaply as Chinese and Japanese labourers.[80] Consequently, as one observer commented, "in some of the eastern provinces they are called 'North American Chinks' because they are willing to work for such low wages."[81] This was confirmed in a letter from the Victoria Whaling Company to the Canadian Department of Immigration Office in Victoria, justifying the hiring of Newfoundlanders:

> The four men who passed before you this morning are going to work on our whaling plants and there are another ten men to arrive, we understand, tomorrow. These men are all experienced in the whaling business and for the last ten years it has been our custom to employ a certain number of Newfoundlanders for our work as they are especially suited … We have no difficulty in getting Oriental labour for our business but we find it very difficult to get suitable white labour in this country and that is our reason for sending back to Newfoundland for men who have had previous experience in our business.[82]

Jas. Murphy and Sons, "Placentia's Fastest Growing Store," handled recruitment in Newfoundland.[83] In April 1916, for example, the company received a "check each month for $175.00 covering $25.00 each from the following men: France [Francis] Mooney, Henery [sic] Mooney, John Green, Wm. Rowe, Ben Mooney, Wm. Judge, John Judge."[84] The amount represented half of their monthly wages and was to be paid to their families in Placentia Bay.

The Newfoundlanders became a noticeable community in Victoria where, according to one company employee, the "old Newfie dialect … could be heard along the Victoria waterfront seventy years ago [1915] among the sealers and whalers wintering here." Most of them had two-year contracts at $50 a month "and found and lodging" and were "mostly cod fishers who [could] turn their hands to most anything, and willing too." They were "clannish as hell, but once you … [get] their confidence and liking you will admire their hardiness … [their] willingness to pitch in … and their wit." It was also this man's opinion that "should a seaman be required on any of the whalers … one of these men would fit in and more than pull his weight … always more than willing to lend a hand wherever needed, I have admired and respected them all of my life."[85]

Transfer of Vessels

The catchers *St Lawrence* and *Sebastian* also came from Newfoundland. *St Lawrence* (ill. 4.11), built in Christiania (Oslo) for Dr Rissmüller's station at Little St Lawrence, continued whaling along the south coast of Newfoundland until late 1906. It was then sold to "parties in Canada"[86] three days after Captain Sprott Balcom visited St John's on 5 October to purchase "one or more of the whalers" from the Newfoundland companies going into receivership.[87] *St Lawrence* left St John's on 23 October 1906, and "making better time than the *Orion*" (the first British Columbia catcher, which had cleared directly for Vancouver Island from Norway in 1905), sailed round Cape Horn and arrived at Victoria on 12 February 1907.[88] The crew began whaling from Page's Lagoon that fall, capturing 245 whales by the end of November worth "over $500,000 – a sum never before approached in the annals of whaling."[89] *St Lawrence* moved north to Kyuquot in the spring and continued to do well – it "broke the world's record"[90] by killing 318 whales. The vessel continued catching until 1930, and was used as a towboat between 1931 and 1948.[91]

Sebastian began its career as *Mic Mac*,[92] built in Christiania in 1904 for the Mic Mac Whaling Company which had just established a station at Dublin

Cove. The catcher whaled for three years with moderate success. Renamed *Sebastian*, it was transferred to Little St Lawrence after its original owners liquidated in 1907. However, Rissmüller had by then decided to close Little St Lawrence and sever connections with Newfoundland whaling. *Sebastian* was hired by the Newfoundland government in 1908 to carry passengers and mail around Fortune Bay[93] before being sold to the Pacific Whaling Company. It reached Victoria in March 1910[94] and hunted for the company until 1916, when as *Saanich* (ill. 5.4) it worked as a tug between 1916 and 1947.[95] The establishment of the Western Whaling Company in 1948 provided it with yet another whaling opportunity as the company's third vessel: the "ninety-three-foot towboat formerly used by the Coastal Towing Company, must have given senior whaleman a sense of déjà vu, for it was none other than the Canadian North Pacific Fisheries chaser boat *Sebastian*, in its livery as the towboat *Saanich*. After a thirty-two-year hiatus it had returned to the whaling business."[96]

The beginning of modern shore-station whaling in British Columbia was but one stage in the global expansion of the modern industry from Norway. Its development coincided with the decline of the Newfoundland and Labrador whale fishery, which provided processing equipment, vessels, and experienced personnel. The British Columbia industry parallelled earlier Newfoundland-Norway linkages in that it initially relied upon long distance east-west relationships. The trained had become the trainers. The adoption by Canada of a modified version of the 1902 Newfoundland whaling act in 1904 also provided further Newfoundland input. Lessons learned on the East Coast contributed to greater longevity of the industry on Canada's West Coast.

CHAPTER 9

Renewal and Revitalization in Newfoundland and Labrador, 1918–51

After almost two decades of uninterrupted whaling, none took place from Newfoundland and Labrador in 1917. A brief resumption occurred at Cape Broyle in 1918 when the station was bought and refurbished by the Newfoundland Shipbuilding Company of Harbour Grace.[1] The company proposed working with Norwegian interests to convert whale oil into margarine and export whale meat to the United States. There "canned whale meat was first introduced into the poorer districts of New York. It has since, however, become a 'delicacy' of uptown New York."[2] Although about one hundred whales were killed, the experiment failed, and the Cape Broyle station closed, this time permanently.

The company's two catchers, *Port Saunders* and *Hawk*, previously owned by the Hawkes Bay Whaling Company and the Cape Broyle Whaling and Trading Company,[3] respectively, were sold to California Sea Products of San Francisco. The vessels subsequently made a significant contribution to the development of Californian whaling, primarily hunting grey whales from stations at Moss Landing and Trinidad. Both were transferred in 1932 to the California Whaling Company and then to San Francisco Sea Products in 1939. The vessels ceased whaling the following year.[4]

The Labrador Whaling and Manufacturing Company was the only company to operate in 1919, even though owner Daniel Ryan had attempted to sell the station after the 1916 season. *Cachalot* took at least forty-one whales from Hawke Harbour (ills. 9.1, 9.2), but this appears not to have been profitable since the factory was once again put up for sale in early 1920.[5]

Optimism for a revival of the industry began to develop in 1920 when it was learned that hunting might be resumed by "new interests" at Cape Broyle. Concerns were raised, however, that experienced Norwegians who had become available after the collapse of their domestic whaling industry could be hired at

Ill. 9.1 Hawke Harbour whaling station, c. 1950 (M. Hennessey)

Ill. 9.2 Flensing crew, Hawke Harbour, c. 1938 (G. Lilly)

cheaper rates than Newfoundlanders. Ryan, an MHA, argued during the debate in the House of Assembly on the "Fish Export Bill" that this practice was nothing new: "Norwegian fishermen could be engaged at much cheaper wages than we are now faced with, that he had been employing Norwegians for his whale factories for sixteen years and always got them at lower rates than our own men, and that there would be greater competition against our fisheries hereafter than ever before in our history."[6] As it happened, whale oil markets continued to be weak, and Cape Broyle remained closed.[7]

Whaling now became confined to opportunistic onshore drives of small species such as potheads, an extremely wasteful traditional practice. Of seventy whales secured in Bonavista Bay in 1922, the "oil and other valuable parts were mostly lost owing to want of facilities for handling."[8] Commercial whaling showed no signs of reviving until early 1923, when a new Norwegian company proposed fitting out a pelagic factory vessel to operate in Newfoundland.[9] With oil prices now improving, this idea was rejected by the Newfoundland government on the grounds that it would threaten plans to reopen the stations at Lark Harbour, Aquaforte, Cape Broyle, Beaverton, and Hawke Harbour.[10] Local whaling was set to enter its second phase of expansion.

THE BRITISH-NORWEGIAN WHALING COMPANY, 1926–34

The British-Norwegian Whaling Co. Ltd was incorporated on 9 October 1926, with most of the three thousand shares owned by Sven Foynd Bruun and Anton von der Lippé of Tønsberg, Norway. Job Bros. of St John's had the largest Newfoundland holding, with sixty shares. A station was built at Grady (fig. 1.9; ills. 9.3, 9.4, 9.5) sixty miles north of Hawke Harbour and was managed originally by a Norwegian, Henry C. Kjøle, who died on his way home to Norway in late 1927.

First established as a Hudson's Bay post, Grady was later used by various Conception Bay fish merchants. The new facility at Watering Cove, Big Grady Island, was equipped with machinery bought in 1927–28 from the Compañía Ballenera Española station at Algeciras, Spain, also owned by Bruun. The purchase included the catchers *Canelinas, Condesa de Moral de Calatrava,* and *Pepita Maura. Condesa* was re-registered as the Newfoundland Whaling Company vessel *Deane. Pepita Maura* was lost when she went ashore near Grady during a gale after a harpooned sperm whale "charged the ship, tore away her rudder, a propeller blade, the stern frame and part of the shoe. The shock was so great that those on board the little ship thought she would be demolished."

Table 9.1
Whale Catch from Grady, 1927–30

Year	*Blue*	*Fin*	*Hump*	*Sei*	*Sperm*	*Total*
1927	1	61	5	1	7	75
1928	0	89	3	0	29	121
1929	–	–	–	–	–	–
1930	0	46	2	0	8	56

Sources: Norsk Hvalfangst-Tidende 18, no. 4 (1929); Hvalfangerføreningen 1930, IWS (various).
Dashes indicate no whaling.

Ill. 9.3 Transport vessel *Herlof* at Grady, 1926 (H. Sorvald)

Ill. 9.4 Grady station under construction, 1926–27 (H. Sorvald)

Left: Ill. 9.5 First whale caught from Grady, 1927 (H. Sorvald)

Although the company caught at least 250 whales from Grady by 1930 (table 9.1) when they temporarily ceased operating, processing was difficult due to the station's location in a "poor harbour, the wharf having to be constructed distant from the plant and connected by a railway." Operations briefly resumed in August 1934, but the station closed at the end of the season.[11]

THE NEWFOUNDLAND WHALING COMPANY, 1923–36

In June 1923 the Labrador Whaling and Manufacturing Company sold *Cachalot* and its stations at Beaverton and Hawke Harbour to Captain Amund Anonsen, a Norwegian who saw a bright future for the Newfoundland and Labrador whaling industry. The Cape Broyle station and the catcher *Cabot* were also bought from the Cape Broyle Whaling and Trading Company by the Newfoundland Whaling Company, which Anonsen incorporated on 17 July 1923.[12]

The new company had begun hunting from Beaverton in June 1923 using *Cabot* (Capt. L. Larsen) and *Cachalot* (Capt. Amundsen).[13] Seventy whales were taken during the first season, yielding 1,600 barrels of oil (table 9.2), despite stormy weather and contrary to mid-season predictions that the catch would not be "as successful as anticipated."[14] Anonsen then decided to close Beaverton and concentrate on whaling from the more productive stations at Hawke Harbour (table 9.3) and Rose-au-Rue (table 9.4).[15] *Cabot* began hunting from Rose-au-Rue in April 1924 before proceeding to Hawke Harbour. *Cachalot* followed later, and hunting ended there in mid-November. The company's second season was a significant improvement over 1923, partly because the Newfoundland government removed export duties on whale and cod oil to encourage further development. Both stations continued to operate in 1925, with *Cachalot* and the new Norwegian-built catcher *Morelos* hunting from Hawke Harbour and *Cabot* from Rose-au-Rue.[16] A small setback was experienced in 1926 when the guano drier at Rose-au-Rue was destroyed by fire.[17]

Whale oil exports from Newfoundland and Labrador during 1927–28 were the most lucrative since the peak seasons of 1904–05. This encouraged the company to buy another catcher, *Rio Sama,* in England to increase catches from Hawke Harbour. The company also began to sell whale meat from Rose-au-Rue as "arctic steak, a palatable and choice meat which is cut from the sirloin of young whales. The meat is instantly brine frozen and wrapped in aluminum foil which gives the surface a neat and efficient protection against light and air."[18] Unfortunately, success was short-lived. Catches declined after 1928, and oil prices again fell due to global oversupply.[19] Consequently, the company stopped operating at the end of the 1930 season, throwing nearly two hundred Newfoundlanders out of work.[20]

The company was liquidated in 1931 and its assets bought for $48,000 by several of the former directors. "Whale farming" was optimistically proposed as part of the new operating strategy.[21] The reconstituted company resumed whaling in the spring of 1933, with *Cachalot, Husvik,* and *Morelos* working from Hawke Harbour. Despite continuing low prices for whale oil, the early part of the season was considered "very successful." Several cargoes of oil were shipped to Denmark and to Lever Brothers in Toronto for use in soap manufacturing.[22] Guano was also extensively advertised locally as a fertilizer: "It is a perfect food for Potatoes, Turnips, and all root crops, also Cabbage and other plants in the garden. It will give you results that are startling. Its high

Table 9.2
Catch and Oil Production, Newfoundland Whaling Co. Ltd, 1923–36

Year	*Blue*	*Fin*	*Hump*	*Sei*	*Sperm*	*Total*	*Oil (Bbls.)*
1923	0	66	3	0	1	70	1,600
1924	12	144	16	0	8	180	5,500
1925	12	270	35	4	10	331	8,400
1926	10	329	18	3	0	360	11,600
1927	15	243	88	9	8	363	14,514
1928	58	358	21	23	48	508	20,580
1929	23	334	11	3	11	382	15,770
1930	23	282	7	1	8	321	13,100
1931	–	–	–	–	–	–	–
1932	–	–	–	–	–	–	–
1933	?	?	?	?	?	?	?
1934	?	?	?	?	?	418	?
1935	4	156	9	13	16	198	7,165
1936	20	146	10	2	14	192	7,186

Sources: Hvalfangerføreningen 1927–30; IWS 1930–52; JNHA (1926; 1927; 1930); Mitchell 1974, 108–69. Cumulative data available do not conform to available individual station data (9.3, 9.4).
Dashes = no whaling; ? = records not available.

Table 9.3
Catch, Newfoundland Whaling Co. Ltd, Hawke Harbour, 1924–36

Year	*Blue*	*Fin*	*Humpback*	*Sei*	*Sperm*	*Total*
1924	?	?	?	?	?	74
1925	?	?	?	?	?	?
1926	?	?	?	?	?	?
1927	3	128	66	1	1	199
1928	7	190	15	5	19	236
1929	14	276	9	0	12	311
1930	21	200	1	0	0	222
1931	–	–	–	–	–	–
1932	–	–	–	–	–	–
1933	?	?	?	?	?	?
1934	?	?	?	?	?	?
1935	1	115	4	13	4	137
1936	15	84	10	1	13	123

Sources: Hvalfangerføreningen (various); IWS (various).
Dashes = no whaling; ? = records not available. For 1935, sixty-one "unidentified" are included here as finwhales.

nitrogen and phosphoric content has won for it great popularity among farmers who used it last season."[23]

Company profits from oil and guano sales were less than hoped for, however, mainly because the factory at Hawke Harbour had to be extensively renovated after lying idle for two years. Stormy weather also reduced late-season catches. As a result the Newfoundland Whaling Company was liquidated for a second time in December 1933.[24]

Despite concerns that stocks had been reduced to such low levels that "in twenty years it may well be that commercial whaling on any large scale will be at an end," the company was reorganized for a third time on 16 January 1934. The most prominent members of the new board of directors were Newfoundland merchants and entrepreneurs C.A. Crosbie, S.W. Moores, C.C. Pratt, and Sir Tasker Cook. Captains Amund Anonsen and Olaf Olsen added a Norwegian presence, with Anonsen serving as operations manager.[25]

Deane, Husvik, and *Morelos* hunted from Hawke Harbour in 1934, with *Cachalot, Rio Sama,* and *Deane* working out of Rose-au-Rue. The company also acquired the station at Grady from the British-Norwegian Whaling Company, helped by the Newfoundland Commission of Government which refunded the annual $1,000 licence fee in exchange for the use of the station for commercial fishing "under the Fishery Development Scheme." Hunting began at Rose-au-Rue in May, Hawke Harbour in July, and Grady in August, with 418 whales being caught before the season ended in late November. The following year the company used Rose-au-Rue and Hawke Harbour only. *Deane* was lost when it

Table 9.4
Catch, Newfoundland Whaling Co. Ltd, Rose-au-Rue, 1924–36

Year	*Blue*	*Fin*	*Humpback*	*Sei*	*Sperm*	*Total*
1924	?	?	?	?	?	?
1925	?	?	?	?	?	?
1926	?	?	?	?	?	?
1927	10	60	8	9	0	87
1928	52	80	3	18	0	153
1929	14	56	2	3	0	75
1930	3	35	4	1	0	43
1931	–	–	–	–	–	–
1932	–	–	–	–	–	–
1933	–	–	–	–	–	–
1934	?	?	?	?	?	?
1935	3	43	5	3	5	59
1936	5	62	1	0	0	68

Sources: Hvalfangerføreningen (various); IWS (various).
Dashes = no whaling; ? = no records available.

ran aground on the way to Hawke Harbour, leaving *Cachalot, Rio Sama, Morelos,* and *Cabot* to continue the hunt. Captain Olsen resigned in January, and his position was assumed by C.A. Crosbie.[26]

Despite their best efforts, the company continued to have financial problems. Before the 1935 season closed, therefore, the board of directors invited the major global whaling enterprise Christian Salvesen of Leith, Scotland, to become an active partner.[27] Salvesen recognized some potential in a Newfoundland and Labrador venture and agreed to provide operating capital, a transport vessel, and catchers in return for control of the company. The board countered with an offer of 2,400 shares, but Salvesen rejected this because it did not give them control "without any fuss." Salvesen required at least 3,301 shares and the resignation of the Newfoundland directors. The demands were met, and Salvesen acquired 3,492 shares at $34,920, retained Anonsen as a director and operations manager, and replaced the board members with Captain Harold Keith Salvesen (later chairman of the board), Theodore Norman Forbes Salvesen, Noel Graham Salvesen, and Iver Ronald Stuart Salvesen.[28] The Salvesen family thus added another company to its substantial whaling interests. Ten experienced Newfoundland workers were immediately despatched to the South Atlantic on the company's vessel *Salvestria*.[29]

The Newfoundland Whaling Company continued to operate throughout the negotiations, using *Cachalot, Morelos,* and *Rio Sama* from Hawke Harbour and Rose-au-Rue. A new initiative was undertaken when whale meat was shipped to Norway to provide feed for fox farms. As well, the Prince Edward Island government advised its fur farmers that Newfoundland whale meat was good for their animals, and the company received an enquiry from the Silver Tip Biscuit Company about the availability of railcar loads. Altogether 192 whales were captured in 1936, slightly fewer than the previous year and significantly down from 1934. Nevertheless, the injection of Norwegian/British capital and experience ensured the continuation of Newfoundland and Labrador shore-station whaling, and catches again approached the peak season of 1904 (fig. 1.10).[30]

The survival of the Newfoundland Whaling Company in one guise or another during the 1920s and '30s was principally due to its use of widely separated stations to extend the hunting season from early April in the south to the end of November at Hawke Harbour on the Labrador coast. The renewed interest in whaling was also prompted by the government's introduction of the 1927 whaling act in an attempt to combat overhunting and to ensure maximum involvement of Newfoundland labour. This fixed the number of stations at six on the island and two in Labrador and required them to be at least seventy-five miles apart, except for those already established at Hawke Harbour and Grady. The licence fee was increased to $1,000 annually, and each station could now be supplied by two catchers rather than one as previously. Only British subjects resident in Newfoundland and Labrador could be employed as ordinary workers, but the nationality of licencees and catcher crews was not restricted. Unfortunately, the number of whales that could be caught from a station continued to be unrestricted. Pelagic factory ships were also prohibited, to encourage the development of land stations and maximize local employment.

THE POLAR WHALING COMPANY, 1939–51

Improved regulations, along with Salvesen's decision to invest significantly in equipment, vessels, and experienced personnel, allowed Newfoundland and Labrador shore-station whaling to continue through the difficult 1930s and into World War II. The industry then entered a postwar expansion phase matched only by the initial period of rapid growth at the beginning of the century (fig. 1.11).

The Newfoundland Whaling Company, now run as a Salvesen subsidiary (Polar Whaling), began hunting from Rose-au-Rue at the beginning of 1937 and later from Hawke Harbour[31] (table 9.5). The catchers *Shila*, *Signa*, *Silva*, and *Spina*[32] were used, and the transport vessel *Sevilla* brought in whaling supplies from Scotland, including time fuses for the harpoon grenades, made in Norway by Ellefsen Tenskruefabbrik.[33] *Shila*, *Signa*, and *Spina* had previously hunted from the station at Saldanha Bay in South Africa operated by Hans Ellefsen Ltd but controlled by Salvesen from 1916–17.[34] The vessels left Cape Town on 26 March 1937 and sailed via Sierra Leone and Madeira to Glasgow where they arrived on 4 May to be refurbished.

Departing Glasgow on 8 June, they reached St John's thirteen days later, bound for Hawke Harbour. *Silva* under Captain Lars Kristensen, meanwhile, sailed from Tønsberg via Ardrossan to St John's, arriving on 17 April en route for Rose-au-Rue to hunt with *Morelos*.[35] The registrations of *Shila*, *Signa*, and *Spina* were transferred to Newfoundland from the Falkland Islands where they had been hunting from Leith Harbour in the Dependency of South Georgia for the South Georgia Company, another Salvesen subsidiary. The crews were largely Norwegians or Scots who worked for Salvesen elsewhere, although locally knowledgeable Newfoundland mates were sometimes hired. Predictably, only highly experienced Norwegians were used as masters and gunners.[36]

The first season for the Newfoundland Whaling/Polar Whaling partnership began in June 1937. The capture of 482 whales by year's end made it "one of the best in the history of whaling this century." Hawke Harbour processed 272 whales, and Rose-au-Rue produced some 3,400 barrels of oil. This success was reported widely and led to renewed speculation locally that the industry would expand and create much-needed employment.[37]

Salvesen's takeover of the Newfoundland Whaling Company was not totally harmonious. Disagreements over operating policy arose,[38] and Salvesen wound up the company in May 1938. Equipment and buildings at Hawke Harbour and Rose-au-Rue, the disused station at Grady, and land at Beaverton, Cape Broyle, and Little Grady Island were put up for sale, along with the old catchers *Morelos*, *Rio Sama*, and *Cachalot* (now described as a freighter).[39] Salvesen's $300,000 bid[40] was accepted on 24 November 1939,[41] and their Newfoundland and Labrador operation was transferred to the Polar Whaling Company.

Table 9.5
Catch, Polar Whaling Co. Ltd, 1937–51

Year	*Blue*	*Fin*	*Humpback*	*Sei*	*Sperm*	*Total*
1937	8	439	9	7	19	482
1938	–	–	–	–	–	–
1939	7	118	4	2	13	144
1940	1	64	7	0	6	78
1941	–	–	–	–	–	–
1942	–	–	–	–	–	–
1943	0	90	4	0	4	98
1944	4	171	6	1	17	199
1945	11	169	9	5	22	216
1946	8	327	5	0	10	350
1947	10	264	4	4	16	298
1948	47	389	13	3	9	461
1949	28	248	8	12	53	349
1950	14	276	7	5	28	330
1951	14	296	20	37	11	378

Source: Vamplew 1975, 286.
Dashes = no whaling.

Anonsen's responsibilities were assumed by Morten Fadum, previously the station manager at Hawke Harbour,[42] and A.H. Murray and Company of St John's became the company's local representative.[43] The final winding-up order for the Newfoundland Whaling Company took effect on 3 June 1943, and the company records were ordered destroyed.[44] Most of the oil produced in 1937 was exported to Rotterdam, where 2,945 tons (2,796 baleen whale oil, 149 sperm) remained in storage during 1938 in the name of the liquidator of the Newfoundland Whaling Company. Salvesen, however, had a "lien upon this oil and owned the storage warrants."[45] About nine hundred tons of oil were also stored at Hawke Harbour.[46]

The Polar Whaling Company did not operate in 1938 because of another oil market collapse. Grade 1 oil declined to just £14 per ton, again primarily because of further oversupply from the expanding and highly successful southern oceans pelagic industry and a record level of oil seed production in the United States.[47] The company resumed hunting on a reduced scale in 1939, with *Shila* and *Signa* working from Hawke Harbour only, taking 144 whales to produce 59,570 barrels of oil and 6,016 sacks of guano. A small number of experienced Newfoundland whalers continued to work for Salvesen in the South Atlantic.[48]

Salvesen's activities in Newfoundland and Labrador were reduced substantially from 1940 when the Ministry of Shipping in London requisitioned their catchers operating from Hawke Harbour for war duties.[49] *Shila*, *Signa*, *Silva*, and *Spina* were destined for coastal defense patrols off Britain since "in moderate waters an average speed of 11 knots could be maintained."[50] Salvesen objected strenously:

> We desire to place on record that this step constitutes a very severe blow to the Polar Whaling Co. Ltd., whose whaling operation depends upon the use of these catchers. The Polar Whaling Co. has already in Labrador a great part of its requirement in the way of whaling stores for the season 1940, and many of these are bound to deteriorate considerably. The Company has also a considerable stock of whale oil and guano for which economical transportation to market was dependant upon whaling operations during 1940. The Company will further be involved in considerable expense in maintaining its whaling stations, of which the expensive plant will deteriorate while idle. We must further record our regret at the decision, in view of the loss of employment to a considerable number of Newfoundlanders for who prospects of alternative employment were nil before the war and are (we under stand) still negligible.[51]

Additionally, the company argued that *Shila*, *Signa*, and *Spina* were particularly unsuitable. They were

> of light construction (lighter we believe than any other catchers built during the last 30 years) – a construction peculiarly suited to whaling certain waters [ice-free waters off South Africa?] only and therefore possibly not suitable to a purpose for which they are being requisitioned – we would urge that you secure confirmation that the Department of Naval Construction gave detailed consideration to this specific point before it was decided that these three catchers should be requisitioned.[52]

Neither the Ministry of Food nor the Ministry of Supply was convinced by the protests, and the transfer proceeded.[53]

The catchers working from Rose-au-Rue were not requisitioned as they were too old (*Rio Sama*, built 1908; *Cachalot*, built 1911; *Morelos*, built 1912) and in poor condition. *Cachalot* had been damaged when it collided with a wharf at Harbour Grace and was put up for sale in May 1941. *Rio Sama* was bought that same year by Johan Borgen, ex-gunner of *Signa*, who wanted to start a "salvage company in Canada."[54]

The war and the loss of their best catchers reduced Salvesen's activities at Hawke Harbour in 1940, and the company stopped hunting until 1943. Rose-au-Rue continued to operate during 1941–42 but as a new venture, Marine Oils Ltd, which shipped oil to England for the manufacture of "butterine."[55] Salvesen returned to Hawke Harbour in 1943, still apparently considering local whaling prospects to be good. A proposal from Messrs Balfour Guthrie of New York to buy the station and two catchers for $19,000 on behalf of private interests who wanted to sell whale oil to the Canadian government was rejected.[56]

The two catchers may have been *Charcot* and *Gun VI*.[57] They were then working out of Leith Harbour, South Georgia, but Salvesen was preparing to transfer them to Labrador to help the company's other catchers supply the high-demand North American oil market.[58] Mindful of earlier requisitions and relocation costs, Salvesen first asked the Admiralty not to target any more of its vessels for war duty until the end of the season. The Admiralty agreed,[59] perhaps influenced by a request from the Ministry of Food that Salvesen be encouraged to continue whaling from Hawke Harbour.

The wartime risks associated with sailing the vessels north made it impossible for the company's manager at Leith Harbour to find a British crew. Instead, non-combatant personnel from the nearby station at Grytviken, belonging to the Compañia Argentina de Pesca, were used. *Charcot* left South Georgia on 19 June 1943 and *Gun VI* on 5 April 1944.[60] The latter arrived in Newfoundland on 30 May after sailing via the west coast of South America and the Panama Canal to avoid a repetition of the "unfortunate delays to *Charcot* which … took 116 days from South Georgia to St. John's Newfoundland on a voyage which under peace conditions would have taken about 48 days. In consequence the vessel arrived in Labrador so late that she was able to catch only one whale before the season closed."[61]

Signa and *Spina* were refurbished in Scotland following their wartime efforts and returned to Hawke Harbour in May 1945 to hunt with *Charcot, Gun VI,* and *Sluga*.[62] *Shila* and *Silva* were unsuitable for the increasingly sophisticated industry and were sold in March 1946, *Shila* to the Faeroe Islands and *Silva* to interests in Haugesund, Norway.[63] *Sluga* was severely damaged in July 1947 running aground near Battle Harbour, Labrador. It was later scrapped at Hawke Harbour (ill. 9.6).[64]

Salvesen's decision to increase activities at Hawke Harbour involved relocating vessels temporarily from South Georgia when they were inactive during the austral winter. Periodically, these included the catchers *Soika*, *Southern Chief*, *Southern Foam*, *Southern Gambler*, *Sposa*, *Sukha*, and *Bouvet III* and the

Ill. 9.6 Hulk of *Sluga*, Hawke Harbour, 1998 (N. Whitman)

Ill. 9.7 Catcher *Sposa* tied up at Conception Harbour, c. 1953 (D. Flynn)

Ill. 9.8 *Sposa* remains, Conception Harbour, 2004 (D. Flynn)

towboats *Southern Hunter*, *Southern Lily*, and *Southern Laurel*.[65] As a result, Salvesen took 461 whales from Hawke Harbour in 1948. With the 295 taken by the Olsen Sealing and Whaling Company from Rose-au-Rue, this produced a record season.[66]

The company also considered relocating its repair base from Harbour Grace to St John's where supplies were more accessible. Additionally, the older *Signa* and *Spina* were sold to "local interests" in early 1948,[67] but *Signa* sprung a leak leaving St John's for Scotland and returned to port to lie idle. She was finally scuttled outside the harbour on 22 November 1953, putting an end to a "career of killing on both sides of the Atlantic during which she had accounted for more than 2,000 whales."[68] Five other catchers were tied up in Conception Harbour (ill. 9.7, 9.8). The remains of *Sposa* are still visible on the beach, and *Soika* was scuttled offshore in 1968. *Sukha* and *Southern Foam* were bought by Dominion Metals for scrap, and *Southern Chief* was sunk by Salvesen in South Georgia in 1961.[69]

The introduction of more modern catchers and increases in annual catches appear to have again reduced local whale stocks. The higher-yielding blue and humpback whales were especially scarce, and catchers often had to make longer voyages into the Gulf of St Lawrence.[70] Besides being more expensive, extended trips delayed processing and so generated lower quality oil.[71] At the end of the war oil demand continued to slump, primarily because margarine consumption was declining as butter became more readily available and affordable. Increased use of synthetic detergent in soap and the production of artificial Vitamin A further reduced demand.[72] However, the industry was helped "by the strong demand for fish meals."[73]

Concerns were now being expressed locally against whaling in general and the Newfoundland industry in particular, especially since it was widely believed to be detrimental to the traditional cod fishery and marine environment.[74] Despite this, whaling was described in 1950 as "one of Newfoundland's most valuable industries ... earning from $800-$1,000 per person"[75] for the approximately 700 men employed from May to November. Some of these men supplemented their incomes by working for Salvesen in the south Atlantic industry. In addition, Polar Whaling employed thirty men at Harbour Grace to maintain the catchers during the winter.[76]

Global whaling faced further problems between 1951 and 1953 as cheaper vegetable oils continued to become more abundant and operating costs increased. High-yielding blue whales were also protected by international agreement from 1952.[77] Nevertheless, Salvesen kept Hawke Harbour working in the hope that markets would return to profitable levels. This did not happen, and the station was closed in October 1951.[78] *Spina* had already been sold to Spain in March 1950, and *Charcot* was eventually purchased by chief gunner Johan Borgen in November 1956.[79] In total, the Polar Whaling Company captured 2,166 whales between 1946 and 1951. Salvesen stopped whaling altogether following the southern voyage of its pelagic factory vessel *Southern Harvester* in 1962–63.[80]

MARINE OILS LTD AND OLSEN WHALING AND SEALING COMPANY

In addition to Salvesen, two smaller companies became involved in local whaling during the war. After leaving the Newfoundland Whaling Company, Captain Olaf Olsen started his own company, Marine Oils, in May 1940. Earlier that year he bought the catcher *Morelos* and the Rose-au-Rue station from the Newfoundland/Polar Whaling companies, then sold them to his newly incorporated company on 10 June 1940 for $7,700 in cash and shares.[81] Marine Oils also attempted to charter *Rio Sama* from the Polar Whaling Company for four to five months work in 1941.[82] Olsen operated Rose-au-Rue from 1942 until 1944 under the management of H.R. Lilly, previously secretary of Newfoundland Whaling, with P.E. Sullivan as assistant manager.[83] The company obtained a concession from the Newfoundland government to build a station at Williamsport[84] (ill. 9.9), and changed the company's name to the Olsen Whaling and Sealing Co. Ltd on 24 October 1944. Olsen initially held the majority of shares, followed by local businessman and ex-Newfoundland Whaling Company director C.A. Crosbie. Crosbie took over the company after Olsen died in February 1945.[85]

Equipment was transferred to Williamsport from Rose-au-Rue. The factory also processed seals and produced herring oil.[86] Rose-au-Rue burned down at the end of the season, and the remnants were put up for sale in November 1945. *Cachalot*, which had been bought from the Newfoundland Whaling Company as a transport vessel, was placed on tender but sank on the way to Trepassey with a cargo of coal. Her crew was rescued by a passing vessel and taken to England.[87]

Hunting began from Williamsport in July 1945 (table 9.6) using *Rio Sama* and the newly overhauled *Morelos* under Captain Johan Borgen. *Rio Sama* was

Table 9.6
Catch and Oil Production, Williamsport, 1945–51

Year	*Blue*	*Fin*	*Hump*	*Sei*	*Sperm*	*Total*	*Whale Oil (Brls.)*	*Sperm Oil (Brls.)*
1945	0	177	0	0	0	177	?	?
1946	3	175	0	0	1	179	?	?
1947	?	?	?	?	?	157	6,490	300
1948	10	277	2	1	5	295	11,367	187
1949	2	177	3	11	?	193	8,506	?
1950	1	133	9	11	1	155	7,200	100
1951	10	187	7	2	1	207	7,555	55

Sources: IWS 1930–52; 1945 and 1947 catches are from *ET*, 26 January 1946; 4 December 1947.
? = No records available.

Ill. 9.9 Flensing, Williamsport, c. 1955 (I. Reid)

abandoned after going aground on 20 January 1946 near St Lawrence while helping *Morelos* tow the Crosbie company's barge *Foundation Jupiter* to their herring reduction plant at Belleoram.[88] The 1945 season was considered a failure despite optimism for the sale of by-products after experiments were conducted by the British Ministry of Food to produce "edible food stuffs."[89]

Morelos was involved in another wrecking incident in 1946 when it helped salvage the British freighter *Empire Thackeray* which ran aground near Lumsden. During the following season it rescued the crew of the British freighter *Langleacraig*, stranded on Great Sacred Island.[90] In 1947 the company replaced *Rio Sama* with *Hektor VII,* renaming it *Olaf Olsen*.[91] A 2,000 ton tanker, *Driller*, was bought to take whale oil to market[92] and *Gun VI* was acquired from the Polar Whaling Company. Optimism for the future of the new station increased when it was rumoured that Lever Bros. might build a "whale oil manufacturing plant" which would need 500,000 tons of oil to "maintain production."[93] A substantial quantity of whale meat was also shipped to the United Kingdom.[94] The catcher *Bronco* under Captain Johan Borgen was chartered in 1948,[95] renamed *Finback,* and used primarily as a towboat.[96] *Finback*, *Gun VI*, *Morelos*, and *Olaf Olsen* continued whaling until October 1951[97] when the station temporarily ceased operations.

By the early 1950s, the company was experiencing financial problems brought on by the depressed price of whale oil.[98] The Williamsport station was remortgaged for $375,000 in April 1950,[99] and consideration was given to converting the Harbour Grace repair facilities into a marine oil hardening plant through a joint venture between the Newfoundland government and "a Swiss-German firm, the Refining and Hardening Company Corona Limited." However, the company decided that "other locations" would be more viable, and the government instead built a boot and shoe factory, also backed by European interests, on the proposed site.[100]

After the station closed, *Finback* was briefly chartered by the Bowater Pulp and Paper Company of Corner Brook for ice-breaking in Humber Arm. The vessel was then sold to Earle Freighting Services of Carbonear for use in the seal hunt.[101] *Olaf Olsen* was chartered by the Newfoundland government in 1956 as a training vessel for engineering students but sank "13 miles northeast of Cape Breton Island when under tow by *Southern Foam* to Sydney."[102] *Gun VI* was advertised for sale as "scrap or possible barge only" in January 1957 but was scuttled the following month off St John's harbour. The company was formally wound down in May 1958.[103]

After a slow and periodic revival in the late 1920s and early 1930s, the presence of Christian Salvesen revitalized the local whaling industry and allowed it to continue through World War II and into the mid-1950s. By that time, relatively low catches, high operating costs, declining postwar oil prices, and improved supplies of vegetable oils made Salvesen's continued involvement in Newfoundland and Labrador whaling impossible. Salvesen did benefit by using the local stations to develop and refine procedures for its larger and more profitable shore and pelagic whaling operations in the South Atlantic.[104] Smaller domestic companies with lower operating costs but less capital also became involved in this renewal. Ultimately, however, they too succumbed to the pressures that had caused Salvesen to withdraw. The Newfoundland and Labrador whaling industry closed, once again.

CHAPTER 10

The Final Demise, 1952–72

Newfoundland and Labrador shore-station whaling entered into its last two decades in 1952. During the first, emphasis was placed on catching small species of whales to provide feed for a growing fur farming industry. Hunting for large species was revived in 1966, primarily in response to the demand for whale meat in Japan. The industry finally ended at the close of the 1972 season when the government of Canada implemented a moratorium on commercial whaling in response to anti-hunting sentiments and growing concern for declining stocks.

SMALL SPECIES WHALING

The two most commonly killed small species in Newfoundland and Labrador throughout the 1950s and '60s were long-finned pilot (pothead) and minke whales. About 55,000 of the former were killed in onshore drives, and at least 223 were taken from shore stations, especially Dildo, after 1948. Particularly large catches during 1951–61 probably reduced the stock and contributed significantly to lower landings. Some 840 minke whales were also caught from shore stations, primarily Dildo.

Arctic Fishery Products Company/Newfoundland Fur Farmers Feed Cooperative

The first attempt to commercially hunt pothead and minke whales on a large scale began with the formation of Arctic Fishery Products Company in 1946.[1] The company intended to convert whale carcasses into animal feed to help ease a shortage on Canadian and local mink and fox farms.[2] Fur farming developed in Newfoundland and Labrador after World War II, when the Commission of

Ill. 10.1 *Arctic Skipper*, Dildo, c. 1955 (I. Reid)

Government introduced a scheme to train twenty ex-servicemen annually for three years from 1946. They were each to receive $500 to help them set up their own operations.[3]

The company began hunting from Dildo in 1946[4] using two small motor boats, *Jigger* and *Dorothy Winter*, as catchers.[5] Shore facilities were enlarged during the winter of 1951, and a wharf was built to accommodate larger vessels, including *Arctic Skipper* (ill. 10.1).[6] The new industry was considered "a great boon to the people in the Trinity [Bay] South area [fig. 1.1], for besides giving quite a bit of employment at the plant [about thirty men], it serves as a market for various kinds of fish and, in the blueberry season, they place large quantities in cold storage."[7]

The company worked closely with the newly established Newfoundland Fur Farmers Feed Cooperative, a consortium of local mink and fox ranchers. Using a government loan, the cooperative built their own processing plant next to the Dildo station (fig. 1.10).[8] Whale oil was exported to the United States, United Kingdom, and the Canada Packers refinery in Montreal.[9] Average yields of 80 pounds of meat and 530 pounds of fat per pothead were reported,[10] which helped convince another company, the Bonavista Cold Storage Company, to build a similar plant at Charleston, Bonavista Bay.[11] Whale meat from this factory was sent to Bonavista for freezing.[12] The facilities at Dildo were again expanded in 1954, and Arctic Fishery Products optimistically proposed placing a "large" net across part of the bay so that a supply of live whales could be maintained for processing.[13]

Potheads were scarce in Trinity Bay during 1954, said to be due to their having "migrated far out into the Atlantic, and finding bait plentiful there, were reluctant to return to the bays."[14] A more realistic explanation, however, was that the large catches of the previous three years had seriously depleted the stock. As a result, Arctic Fishery Products expanded its hunting into Bonavista Bay where large schools were still reported.[15] In April 1955 the Newfoundland government also became directly involved in whaling, fitting out larger vessels to hunt potheads and minkes further offshore in an attempt to ensure a supply of feed for the fur farms. One such vessel was *Matthew II,* originally an "experimental off-shore fishing boat … designed for Danish seining and other modern fishing methods."[16] The availability of an apparently abundant and inexpensive source of food provided an incentive for mainland Canadian mink and fox breeders to transfer their operations to Newfoundland.[17] The involvement of government-sponsored vessels helped increase the minke catch from thirteen to fifty-seven in 1955–56 (appendix 1).

Despite the growth of this new industry in rural Newfoundland, concern was voiced in the press that increased hunting pressure would seriously deplete stocks. Many also considered it inappropriate for the government to subsidize fur farming, an economic sector heavily dependent on volatile markets with increasing competition from synthetic substitutes.[18] As had often been the case since commercial whaling began locally in 1898, fears were again expressed that the dumping of processing waste and offal at sea would pose a health risk for residents of nearby communities.[19]

Premier Joseph R. Smallwood ignored these concerns and continued to promote the fur industry, including the introduction of government incentives to entice more mainland farmers to Newfoundland.[20] The industry, however, did not live up to Smallwood's high expectations. In 1957, for example, three of the five fur farms operating at New Harbour, Trinity Bay, closed.

Operating costs increased as it became more difficult for the cooperative at Dildo to deliver sufficient quantities of low-priced feed[21] because of whale stock depletion.[22] Some farmers began to mix their own feed. One particular recipe called for "five percent pothead meat, 5 percent minke whale meat, 10 percent horsemeat, 10 percent chicken by-product, 10 percent beef tripe, two percent whale liver, five percent beef liver, two percent brewers yeast, 30 percent fish, 21 percent cereal."[23]

Additionally, oil prices increased after the war, and greater emphasis was placed on this aspect of the industry throughout the late 1950s. With grade 1 oil fetching £89 in 1956,[24] a 4,000-gallon storage tank was built at Dildo in 1957 as a substitute for several smaller tanks,[25] and a conveyor belt was added to speed up the movement of blubber, which had previously been hand-carried in buckets from the flensing plan (ill. 10.2) to the digestors[26] (ills. 10.3; 10.4).

Federal government scientists now warned that the ultimate dependence of fur farming on fluctuating numbers of squid, the primary food of pothead whales, would make it difficult to sustain the industry.[27] Nevertheless, they renewed the whaling licence issued to the "Provincial Department of Fisheries

Ill. 10.2 Sperm whale on flensing plan (slipway), Dildo, c. 1967 (W. Hoek)

Ill. 10.3 Ramp from plan to digestors, Dildo, c. 1967 (W. Hoek)

Ill. 10.4 Whale meat and bone being lowered into digestor, Dildo, c. 1967 (W. Hoek)

on behalf of the Newfoundland Fur Farmers Feed Cooperative."[28] Newfoundland continued to be promoted as a prime site for mink and fox farming because of the availability of good, cheap food and the absence of disease.[29] A general air of optimism and prosperity prevailed, particularly after Arctic Fishery Products Company increased its whaling effort in the spring of 1959. A new meat-processing plant was built at Southern Harbour, and further improvements were made at Dildo. *Arctic Skipper* (Capt. Ivor Iversen) and *Arctic Venture* (Capt. Clarence George) (ill. 10.5) hunted in Trinity and Bonavista bays, whilst *Shirley and Gladys* (Capt. Joshua Reid) and *Erikson* (Capt. Lewis Reid) were used primarily to drive potheads into Trinity Bay.[30] The latter was built at Dildo in 1960 with assistance from the provincial government.[31]

Arctic Fishery Products Company and the cooperative processed 1,000,000 pounds of meat from eighteen minke whales and 309,000 pounds from 546 potheads in 1959, enough to feed the sixty thousand mink being raised on sixty local ranches.[32] By 1962, however, the scarcity of pothead whales renewed fears that feed supplies could not be guaranteed.[33] Local whaling was now "virtually non-existent," the following year being the "worst on record."[34] As a result, fur farming declined to about twenty ranches in 1964.[35] The operators hoped that more minke whales could be caught to offset the low availability of potheads.[36] This did not happen, and whale meat was imported for resale at 12 cents a pound compared to 4–6 cents a pound for the local feed.[37] A further blow to fur farming came in 1965 when more than four hundred mink valued at $20–$25 apiece were killed by a viral infection.[38] Food costs continued to rise, and by 1974 small-scale breeders were flooding global markets and lowering pelt prices.[39] Mink farming in Newfoundland therefore declined again.[40] With

Ill. 10.5
Arctic Venture,
Dildo, c. 1955
(V. George)

both whaling and fur production on the wane, Arctic Fishery Products Company put *Arctic Skipper* up for sale in early 1968.[41] She was bought by Captain Henry Mahle and renamed *Iversen* after her skipper.[42] The following year, the Dildo factory began processing seal skins for the Carino Company of Halifax, a subsidiary of G.C. Reiber, Bergen, Norway.[43]

The reduction in pothead stocks was primarily caused by onshore drives. Less than three hundred of those caught between 1941 and 1971 were taken by Dildo catchers, compared to some 55,000 slaughtered on local beaches (ills. 1.1, 10.6, 10.7).[44] Their blubber and meat were also processed at Dildo, some of it for human consumption as "arctic steak."[45] Fishermen received about $3–$4 for the blubber and $10 for the meat.[46] Although the onshore drive generated considerable ethical debate,[47] construction of the processing facilities at Dildo ensured its continuation.[48]

Ill. 10.6 Traditional "pothead drive," Chapel Arm, Trinity Bay, 1955 (R.E. Eshelman)

Ill. 10.7 Flensing, traditional "pothead drive," Chapel Arm, Trinity Bay, 1955 (R.E. Eshelman)

LARGE SPECIES WHALING

The pursuit of small species to supply the expanding fur industry dominated postwar commercial whaling in Newfoundland and Labrador. Between 1951 and 1965 only twenty-six large whales were taken from Dildo (table 10.1) by *Arctic Skipper* and *Arctic Venture* for Arctic Fishery Products. This low catch was partially influenced by limited processing capacity,[49] but also because the seasonal occurrence of large numbers of potheads made "pursuit of the larger species of whale unnecessary, and these boats are now being used to assist the smaller boats in the game of beaching the potheads."[50]

The Hawke Harbour Whaling Company, 1956–59

The large-species sector of the industry received a boost with the reopening in 1956 of the station at Hawke Harbour, which had closed five years earlier when Christian Salvesen Company withdrew from Newfoundland and Labrador.[51] Captain Johann Borgen was the major shareholder in the newly established Hawke Harbour Whaling Company.[52] The ex-Salvesen/Polar Whaling Company catchers *Sposa* and *Southern Foam* were used to supply the station until it was destroyed by fire on 12 September 1959.[53] During this period, 150 whales were processed (table 10.2).[54] *Southern Foam* was damaged later in the year when she ran aground, but was repaired and tied up in Conception Harbour with *Sposa*.[55] They were not used again.[56] A new company, Newfoundland American Whaling Production Company, was unable to raise sufficient capital to rebuild the station.[57]

Further Connections with British Columbia

Large-species whaling in Newfoundland and Labrador was limited between 1951 and 1956, leading to further local involvement in the British Columbia

Table 10.1
Large Species Catch, *Dildo*, 1951–65

Year	*Fin*	*Sperm*	*Hump*	*Sei*	*Total*
1951			2		2
1952	1		1		2
1953	1				1
1955	2				2
1957	1				1
1958	3		1		4
1960	1	1			2
1961				1	1
1964	1			1	2
1965	6		1	2	9
Total	16	1	5	4	26

Source: IWS

industry. Experienced personnel moved west to work seasonally at the Coal Harbour station on Vancouver Island (fig. 8.1) operated by the Western Whaling Company for British Columbia Packers. Recruitment began in 1955,[58] largely through the intermediacy of Captains Arne Borgen and Wallace Anstey, both then whaling for B.C. Packers.[59] The scheme was supported by the federal and provincial governments.[60]

Despite the availability of steady seasonal employment, the Newfoundland whalers, as members of the United Fishermen and Allied Workers union (UFAW), decided not to participate in the 1959 hunt on the West Coast unless a new contract was introduced to replace the 1956 agreement under which they were operating. Specific demands included increased monthly wages,[61] a change in bonus payments from a system based on rank to a flat rate of $2 per whale, and paid return travel between Newfoundland and British Columbia rather than one way only. A settlement was reached that included a monthly wage increase of $20 for 1959 and $15 for 1960.[62] Unfortunately, the Newfoundlanders were not welcomed by many of the local UFAW members, who expressed the opinion that unemployed British Columbia seamen should receive hiring preference.[63] Despite this, the eastern whalers continued to be valued employees until the company stopped whaling at the end of the 1960 season.[64]

Another Newfoundland contingent was recruited in 1962 to work at Coal Harbour for the Western Canada Whaling Company, a consortium of B.C. Packers and the Japanese company, Taiyo Fisheries Ltd. The main objective was to produce frozen whale meat,[65] then in high demand in Japan.[66] Shipping was not economically viable, however, "due to currency controls,"[67] and the meat was instead sold at significantly lower prices to United States mink ranchers. As a result, disgruntled Newfoundland workers received only $2,000 per season and had to pay $500 towards their travel costs.[68] This situation was redressed in 1964,[69] and they continued to participate in British Columbia whaling until it ended in 1967.[70]

The Mid-1960s: The Last Revival

The decline of British Columbia whaling in the 1960s coincided with a resurgence of the industry in Newfoundland and Labrador.

Table 10.2
Catch and Oil Production, Hawke Harbour Whaling Co. Ltd, 1956–59

Year	*Fin*	*Hump*	*Sei*	*Sperm*	*Total*	*Whale Oil (Brls.)*	*Sperm Oil (Brls.)*
1956	7	0	2	13	22	269	598
1957	23	0	5	14	42	960	760
1958	55	4	0	7	66	1,246	285
1959	14	0	5	1	20	640	82
Total	99	4	12	35	150	3,115	1,725

Source: IWS

Ill. 10.8 Preparing whale meat for freezing, Dildo, c. 1970 (Whaling and Sealing Museum, South Dildo)

Table 10.3
Large Species Catch, Dildo, 1966–72

Year	*Fin*	*Sperm*	*Hump*	*Sei*	*Total*
1966	164	2	0	0	166
1967	174	0	0	4	178
1968	219	0	0	4	223
1969	188	5	1	3	197
1970	181	2	9	1	193
1971	117	0	10	0	127
1972	115	2	0	0	117
Total	1,158	11	20	12	1,201

Source: IWS

Table 10.4
Large Species Catch, Atlantic Whaling Co. Ltd, Williamsport, 1967–72

Year	*Fin*	*Hump*	*Sei*	*Total*
1967	262	0	3	265
1968	219	0	0	219
1969	188	4	0	192
1970	227	5	0	232
1971	184	6	0	190
1972	150	0	1	151
Total	1,230	15	4	1,249

Source: IWS

Ill. 10.9 *Westwhale 8*, Dildo, c. 1970 (W. Hoek)

Whaling from Dildo received a boost in 1966 (table 10.3) when a new enterprise, the East Coast Whaling Company, began exploratory hunting in conjunction with Taiyo Fisheries.[71] Their catcher *Kyo Maru 17* operated from mid-September to mid-November that year within 140 miles of the station so that carcasses could be quickly processed to produce optimum quality meat for human consumption (ill. 10.9).[72] The vessel was replaced by *Westwhale 8* (ill. 10.8), bought from the Western Whaling Company, and sailed to Dildo by Captain Arne Borgen and a crew of expatriate Newfoundlanders. She hunted locally until Canadian whaling ended.[73] The company also used *R.D. Evans* and *Happy Adventure*. The former was a stern trawler with a 50mm harpoon gun mounted seasonally, and the latter was built by Captain Clarence George. He also built *Trinity Star* in 1972, but the industry closed before she could be brought into service.[74] Frozen meat was exported to Japan for human consumption and to the United States for animal feed. This market closed in 1970 when the United States government banned the import of whale products.[75]

Taiyo Fisheries and the Fishery Products Company also entered into a partnership in October 1966 as the Atlantic Whaling Company,[76] and re-opened the station at Williamsport. Thirteen of the eighty shore employees were Japanese, including the station manager. Their catcher, *Fumi Maru 15,* carried a fourteen-man crew, seven of whom were Newfoundlanders. As with other companies, the scale of the operation was dictated by catch quotas imposed by the Government of Canada.[77] Whaling ended at Williamsport in 1972 (table 10.4; ills. 9.7, 9.8) following a decision by company shareholders to build a new factory at a more accessible location near St Anthony. This plant was not completed due to the imposition of the moratorium. The last cetaceans were killed by Newfoundland and Labrador commercial shore-station whalers seventy-four years after the first has been taken off Baccalieu Island in Conception Bay.

Epilogue

GENERAL SUMMARY

The modern Newfoundland and Labrador shore-station whaling industry was highly cyclical, with well-defined catch peaks in four periods: 1903–05, 1925–30, 1945–51, and 1966–72. Contributing factors to the declines included ineffective management, stock overexploitation, low capital, high operating costs, prohibitive expenditures on new stations and catchers, volatile markets for oil, bone, and by-products, and an initial dependence on the more highly paid Norwegians.

The Norwegian presence was significant throughout the industry, especially in its formative stage. Partnerships developed between Norwegian and Newfoundland entrepreneurs, the first, the Cabot Steam Whaling Company, established in 1896. This was a direct result of overexploitation of whale stocks off Finnmark and the need for Norwegian whalers to find new hunting grounds to recoup their investments. The relatively untouched whale stocks in the western North Atlantic and the fortuitous presence of a Norwegian as the first superintendent of Fisheries for Newfoundland and Labrador provided the opportunity in 1898 to transfer Norwegian whaling expertise to the colony and begin this new phase of global whaling.

Cabot Steam Whaling demonstrated that whaling could be successful in Newfoundland waters. This precipitated the development of more companies and a rapid expansion of the industry. Faced with concern that the pace of growth would deplete stocks, reduce profits, and destroy the industry, the Newfoundland government passed the first act to regulate whaling operations in 1902. This act became a model for control legislation introduced by other jurisdictions internationally in the early twentieth century.

Unfortunately the act did not provide the necessary conservation framework. It did not set limits on the number of whaling licences issued, for example, or on an annual quota per licence. Catches, production, profits, and competition therefore continued to increase, especially in the 1904–05 seasons. As a result, the predicted overexploitation occurred, revenues declined, stations closed, and the first cycle came to an end in 1916.

Contrary to expectations, the new industry had little impact on the socio-economic conditions of most rural residents, many of whom had already experienced a serious decline in the cod and seal fisheries, their economic mainstays. By local standards, however, those engaged by the whaling companies had benefited greatly, even though they were generally employed in menial, low-paying processing jobs onshore, whereas the higher-paying positions on the catchers were filled by experienced Norwegians. More lucrative whaling opportunities became available after 1905 when experienced personnel were able to transfer to the growing whaling industry of British Columbia. While also relying on Norwegian expertise, the B.C. venture would probably have been less successful if technology and experience had not been brought in from Newfoundland.

The Newfoundland industry did not fully resume until 1923. Substantially improved oil prices, an apparent abundance of whales, and export incentives from the government led to the formation of the Newfoundland Whaling Company, which, with a new British-Norwegian whaling company operating from Grady in Labrador, briefly revived the industry. However, catches and oil prices again declined, and the industry closed for the second time. It did not reopen until 1934 when the Newfoundland Whaling Company was reconstituted. A diminished supply of whales and deteriorating markets quickly brought about another temporary cessation.

Some measure of longevity occurred when Christian Salvesen bought the Newfoundland Whaling Company and operated it as the Polar Whaling Company into the early 1950s. Economic conditions caused Salvesen to also withdraw, along with Olsen Whaling and Sealing, the only other company then operating. Interest was renewed in the early 1960s to provide meat and by-products from minke and pothead whales for a growing fur-farming industry. This led to a revived hunt for large species in the mid-1960s, in part to supply the demand for meat in Japan. After more than four centuries, commercial shore-station whaling in Newfoundland and Labrador finally closed in December 1972 when the Canadian government implemented the moratorium.

During the modern phase from 1898, twenty-one stations operating for varying lengths of time processed almost twenty thousand whales. Today all the facilities have been demolished, or are in the latter stages of collapse (ills. 11.1, 11.2, 11.3). A few items of rusting machinery, remnants of concrete slipways, and grassy clearings can still be found, but in general there is little physical evidence to indicate the nature and extent of these stations (ills. 11.4, 11.5). The whaling industry of Newfoundland and Labrador, like most others throughout the world, is now of historical interest only (ills. 11.6, 11.7).

Ill. 11.1 Williamsport station, 1994 (C.W. Sanger)

Ill. 11.2 Rusting whale oil and fuel tanks, Williamsport, 1994 (C.W. Sanger)

Ill. 11.3. Main wharf and rusting boilers, Williamsport, 1994 (C.W. Sanger)

Ill. 11.4 Site of former Trinity whaling station, 1990; for comparison, see ill. 5.1 (A.B. Dickinson)

Ill. 11.5 Site of former Snook's Arm whaling station, 1990; for comparison, see ills. 2.2, 2.4, 3.6, 3.8 (A.B. Dickinson)

Ill. 11.6 Whaling and Sealing Museum, South Dildo, 1991 (C.W. Sanger)

Ill. 11.7 Post-industry decorative use of whaling equipment and materials (55 mm. harpoon gun and whale bones), private garden, Dildo, 2004 (A.B. Dickinson)

AFTERMATH

Commercial shore-station whaling in Newfoundland and Labrador operated during its early years under ineffectual management procedures introduced in the 1902 act. Although these procedures were revised and strengthened in 1927, an absence of objective information on species biology and the size and distribution of individual stocks continued to hinder control.

Following confederation in 1949, the Canadian government became responsible for managing resources in the waters off Newfoundland and Labrador. Canada had approved the 1946 *International Convention for the Regulation of Whaling* (ICRW) and joined the International Whaling Commission (IWC) in

1949.[1] As a result it became necessary to determine the abundance and distribution of whales in Canadian territorial seas. The beginning of whaling from Blandford, Nova Scotia, in 1964 and the resumption of fin whaling from Dildo and Williamsport in 1966 gave the Fisheries Research Board of Canada an opportunity to monitor commercial catches and obtain catch-effort and life-history data for the western North Atlantic. They also carried out tagging and census cruises and began tracking humpback distribution by recording the fluke patterns of individual whales.[2] These efforts were seriously reduced after 1972, when the industry closed.[3] However, the IWC and the U.S. National Marine Fisheries Service (NMFS) continue to provide abundance estimates and distribution data.[4] The conservation status of various species is also categorized by the Committee on the Status of Endangered Wildlife in Canada (COSEWIC).[5]

The impact of commercial whaling on fin whales in the western North Atlantic epitomizes the conservation issues involved. There may have been fifty thousand of this species distributed throughout the North Atlantic before the industry began, with one stock concentrating for summer feeding off Newfoundland and Labrador. Much-reduced catches of increasingly smaller whales between 1951 and 1965 and a declining catch per unit effort indicated overexploitation. In an attempt to redress the decline, the IWC set annual quotas for each station between 1967 and 1972 (table 11.1), by which time the local industry was "close to the point where a profit could no longer be made."[6] The abundance estimate was also revised, from 4,500 in 1967 to 2,370 in 1972.[7]

In case whaling did resume after 1972, the IWC continued to allocate a western North Atlantic quota (ninety-five in 1973) until 1982 (ninety),[8] when it announced a ten-year global moratorium on commercial whaling.[9] Since the whaling industry in Atlantic Canada depended for its existence on the availability of fin whales, and with 95 per cent of the quota being taken for little profit, economic factors would in any case have brought the industry to a close.

The IWC estimates the present North Atlantic fin whale stock at 47,300, close to the presumed pre-whaling size. The NFMS conservatively suggests a minimum of 2,814 whales in the western North Atlantic, perhaps increasing in number. Fin whales, however, continue to be listed under the U.S. Endangered Species Act, and are "of special concern" to COSEWIC.[10] Although the IWC protects the Nova Scotia stock from the unlikely prospect of further hunting, fin whales off Newfoundland and Labrador are not accorded the same treatment.[11]

Catch data also indicate that blue whales were overexploited. They were not caught from Newfoundland and Labrador after 1951 and were protected by the IWC in Arctic and North Atlantic waters from 1955. There are now between 1,100 and 1,500 in the North Atlantic, with about 308 (NFMS) in western waters compared to a possible pre-whaling stock of 1,500. Population trends cannot be determined because of insufficient data, and blue whales are listed under the U.S. Endangered Species Act. COSEWIC also considers them to be endangered.[12]

The IWC protected humpbacks in the North Atlantic in 1955 but allowed forty-one to be taken under scientific permit from Canadian East Coast stations between 1969 and 1971.[13] The IWC estimate a current North Atlantic stock of 11,570. Although apparently increasing in number, humpbacks continue to be listed under the U.S. Endangered Species Act. However, COSEWIC considers that the western North Atlantic stock is not at risk.[14]

Pothead or pilot whale numbers were also seriously reduced, especially by the large onshore drives from 1947 when there were probably 50–60,000 off eastern Newfoundland. An aerial survey in 1980 estimated 14,000, and the NFSM suggests a current minimum of 11,343 in the western North Atlantic. Although they have not recovered to their estimated pre-exploitation number, potheads are not listed under the U.S. Endangered Species Act, and COSEWIC does not consider them at risk.[15] However, the recent absence of large strandings suggests that they are now scarce in some locations.[16]

While there are probably less than three thousand sei whales in the North Atlantic, they are also not listed by the U.S. Endangered Species Act or COSEWIC. Those off Nova Scotia are protected by the IWC, whereas the Newfoundland and Labrador stock is not.[17] At least 3,505 sperm whales (NFMS) may frequent the western North Atlantic. Although data are insufficient to determine the population trend, they are not considered at risk by COSEWIC, and the IWC does not identify them as a protected stock.[18] Some 3,515 minke whales (NFMS) are also estimated off Canada's East Coast, with a further 149,000 (IWC) elsewhere in the North Atlantic. They are not listed under the U.S. Endangered Species Act or by COSEWIC.[19]

Although overexploitation and economic factors contributed significantly to the periodic declines of the Newfoundland and Labrador whaling industry, its ultimate ending resulted from the introduction of policies to protect whales globally. In 1972 the United Nations Conference on the Human Environment passed a resolution calling for a ten-year ban on commercial whaling,[20] and in July of the same year the United States enacted the *Marine Mammals Protection Act* prohibiting commercial whaling in U.S. waters by U.S. citizens, and the

Table 11.1
Fin Whale Catches and Quotas, Western North Atlantic, 1967–72

Station	1967	1968	1969	1970	1971	1972	Total	c%q
Blandford	c 318 q 300	q 262 c 262	c 157 q 224	q 170 c 150	c 117 q 110	q 95 c 95	c 1,119 q 1,141	98
Dildo	c 174 q 250	q 220 c 219	c 188 q 188	q 181 c 225	c 117 q 160	q 115 c 115	c 995 q 1,157	86
Williamsport	c 271 q 250	q 219 c 219	c 188 q 188	q 225 c 225	c 184 q 200	q 150 c 150	c 1,237 q 1,232	100
Total	c 763 q 800	q 701 c 700	c 533 q 600	q 576 c 600	c 418 q 470	q 360 c 360	c 3,351 q 3,530	95

Sources: K.R Allen 1971; Mitchell 1974.
c = catches; q = quotas.

importation of whale products.[21] The Canadian government followed by placing a moratorium on the national industry on 22 December 1972. Further international support to end commercial whaling came in 1973 when the United Nations–sponsored *Convention on International Trade in Endangered Species of Wild Fauna and Flora* (CITES) outlawed trade in whale products.[22] In 1982 the IWC implemented the 1972 UN resolution by placing a global moratorium on commercial whaling, effective in 1986 for coastal stations.[23] Canada also finally legislated against commercial whaling in its waters in June 1982, when it introduced *Cetacean Protection Regulations* as an amendment to the Fisheries Act. Specific provisions continue to permit hunting by indigenous peoples.[24]

Commercial whaling, however, still continues globally, either regulated by the IWC or not. Japan implements the major hunt, for minke whales in southern oceans under the guise of "scientific whaling,"[25] and is the primary market for meat. Iceland, Norway, and the Faeroe Islands continue the hunt in the North Atlantic. Iceland has a long whaling history, taking some 35,200 whales since data began to be collected in 1883. The IWC established a North Atlantic quota for minke whales in 1973, but the Icelandic hunt for them ended temporarily in 1985 after the IWC moratorium was announced. Fin and sei whales were also taken for "scientific purposes" (Article VIII, ICRW) from 1986 until 1989 under quotas and guidelines established by the national government. Iceland resigned from the IWC in 1992 but was re-admitted in October 2002 on the condition that commercial whaling would not resume before 2006. However, the hunt began earlier through the "scientific whaling" loophole. Three vessels began catching minke whales in mid-August 2003, taking thirty-six from a domestically established quota of thirty-eight. Iceland considers that an estimated local stock of 43,000 minke whales can support a hunt and intends to continue whaling, primarily to export meat to Japan.[26]

Norway also ceased coastal whaling in 1986 when the IWC moratorium took effect. The Norwegian government subsequently objected as allowed under IWC regulations, and commercial hunting resumed in 1993. Only minke were taken from 1977. Annual quotas are set by Norwegian authorities based on the Revised Management Scheme developed by the IWC to regulate any future whaling that might occur under its jurisdiction. The quota of 711 for 2003 was based on a central and northeast Atlantic stock minimally estimated at 120,000. Meat is sold on the domestic market, and the legal export of products not used locally began in July 2002 with the shipment of meat to Iceland. On 27 May 2003, Japan announced that it would begin importing whale products from Norway, which has a CITES reservation permitting this. The Faeroese continue their traditional government-regulated onshore drive of pothead whales. An average of 950 were killed each year during the 1990s, the annual yield of 500 tons of meat being approximately 30 per cent of that produced locally. Minke whale meat began to be imported from Norway in 2003.[27]

Aboriginal subsistence whaling quotas are set by the IWC. In the North Atlantic the Bequians of St Vincent and the Grenadines are allowed to take up to twenty humpbacks for local consumption between 2003 and 2007. Two were taken in 2002. Annual strike limits of 175 minke and nineteen fin exist for residents of west Greenland, and twelve minke for east Greenland. In 2002, thirteen fin and 139 minke were taken by west Greenland and ten minke by east Greenland. In Canada the signing of the Nunavut Land Claims Agreement in 1993 between the territory and the Government of Canada allowed the Nunavut Wildlife Management Board to authorize the Inuit of Hudson Bay to take one bowhead whale every two years for food. One was killed in 2002.[28]

Canada announced in 1981 that it was leaving the IWC since it "no longer has any interest in the whaling industry or in any related activities of the IWC." It continues to hold observer status,[29] however, and also membership in the North Atlantic Marine Mammal Commission (NAMMCO) established by Norway, Iceland, Greenland, and the Faeroe Islands in 1992 to promote conservation, scientific study, and sustainable utilization "with due regard to the needs of coastal communities and indigenous peoples."[30]

Although no longer hunted locally, whales are of growing importance to the socioeconomic fabric of coastal Newfoundland and Labrador. Tourism is being increasingly promoted in the province and includes a significant whale-watching industry. With some fifty commercial whale-watching operations focusing on humpback, fin, and minke, the proximity of people and boats provides the potential for increased stress on the animals. The IWC has recognized the need for precautionary measures globally, and initiatives to regulate the Canadian industry have been recommended to both levels of government.[31] Until these are implemented, whales that seasonally migrate through the waters off Newfoundland and Labrador may continue to be adversely affected by human activity.

Appendices

APPENDIX I

Shore-Station Whale Catch Species Composition, Newfoundland and Labrador, 1898–1972								
Year	*Blue*	*Fin*	*Hump*	*Sei*	*Sperm*	*Minke*	*Other*	*Total*
1898							91	91
1899							109	109
1900							190	190
1901							258	258
1902							450	450
1903	225	345	287				1	858
1904	264	690	281	39	1			1275
1905	265	460	161	2	6			894
1906	96	272	58	2	1			429
1907							481	481
1908	36	345	24		1			406
1909	48	385	79	2	4			518
1910	33	295	56					384
1911	34	249	26		7		19	335
1912	60	202	22		5			289
1913	12	165	8	1	9		27	222
1914	5	142	13		1			161
1915		115	5				19	139
1918							101	101
1923		66	3		1			70
1924	12	144	16		8			180
1925	12	270	35	4	10			331
1926	10	329	18	3				360
1927	14	249	79	11	8			361

Year	Blue	Fin	Hump	Sei	Sperm	Minke	Other	Total
1928	57	359	21	23	48			508
1929	28	332	11	3	12			386
1930	24	281	7	1	8			321
1935	4	156	9	16	19			196
1936	20	146	11	1	13			191
1937	8	439	9	7	19		1	483
1939	7	118	4	2	13			144
1940	1	64	7		6			78
1941	2	65	3	2				72
1942	4	62	1	4				71
1943	1	141	6		4		98	250
1944	5	231	10	1	17			264
1945	11	346	9	5	22			393
1946	11	502	5		11			529
1947	14	413	6	4	18		157	612
1948	57	670	15	4	14	38	9	807
1949	30	425	11	23	53	38		580
1950	16	408	16	16	29	3	172	660
1951	24	483	29	39	12	17	62	666
1952		1	1			20		22
1953		1				32	1	34
1954						13	20	33
1955		2				57	1	60
1956		7		2	13	36	32	90
1957		23		5	14	37	19	98
1958		55	4		7	42		108
1959		14		5	1	18		38
1960		1			1	11		13
1961				1		22		23
1962						45		45
1963					1	18		19
1964		1		1		35		37
1965		6	1	2		29		38
1966		164			2	28		194
1967		436		7		25		468
1968		438		4		50		492
1969		376	5	3	5	50		439
1970		408	14	1	2	86	7	518
1971		301	16			73	6	396
1972		265			2	97	2	366
Total	1450	12863	1402	246	418	920	2333	19632

Sources: Newfoundland, *Annual [Fisheries] Reports*; IWS; *Journal of the Newfoundland House of Assembly*; Mitchell 1974a, 1975a; Mitchell and Kozicki 1975; Mitchell and Reeves 1981, 1983; newspapers (various)
Note: Differences exist between sources, especially for smaller species. In 1974, for example, IWS reported catches for only the Polar Whaling Company (Hawke Harbour); however, the Olsen Whaling and Sealing Company Ltd killed 157 that year (*ET*, 4 December 1947). These are included above as "other" since species composition was not identified.

APPENDIX 2

Whaling Companies, Stations, Operating Years, and Vessels in Newfoundland and Labrador, 1898–1972

Company	*Stations*	*Operating Years*	*Catchers*
Cabot Steam Whaling	Snook's Arm	1898–1905, 1907–14	*Cabot*, *Viking*
	Balaena	1899–1917	
Newfoundland Steam Whaling (= Rose-au-Rue Whaling, 1910)	Reuben's Cove (Chaleur Bay)	1901–07 (?)	*Puma*, *Lynx*
	Rose-au-Rue	1902–05	
	Beaverton	1904–07 (?)	
Cape Broyle Whaling and Trading	Cape Broyle	1903–08, 1904–12	*Falcon* (*Viking*)
	Cape Charles	1904–10	*Hawk*
Aquaforte Whaling (Ellefsen)	Aquaforte	1902–07	*Humber*
St. Mary's Whaling and Trading	Riverhead	1904–06	*Avalon*
St. Lawrence Whaling (Rissmüller)	Little St Lawrence	1903–07	*St Lawrence*
Mic Mac Whaling (= Phoenix Whaling, 1905)	Dublin Cove	1904–14	*Mic Mac*
	L'Anse-au-Loup	1904–05	(*Sebastian*, 1907), *Avalon*
Labrador Whaling and Manufacturing	Hawke Harbour	1905–16	*Cachalot* (2, first lost 1910)
Atlantic Whaling and Manufacturing (= Phoenix Whaling, 1905)	Trinity	1904–05, 1907–14	*Fin*
	Safe Harbour	1904–06	*Hump*
Harbour Grace Whaling (later Newfoundland Shipbuilding)	Harbour Grace	1905–06 1910–13 (as Hawke's Bay Whaling Co. Ltd.)	*Baccalieu*
	Cape Broyle	1918	*Port Saunders* *Hawk*
Hawke's Bay Whaling	Hawke's Bay (Port Saunders)	1905–11	*Port Saunders*
	Harbour Grace (with Harbour Grace Whaling Co)	1910–13	
Newfoundland Whaling (reorganized 1931, 1934)	Beaverton	1923	*Cabot*, *Cachalot*,
	Hawke Harbour	1924–30, 1933, 1934, 1936	*Morelos*, *Rio Sama*
	Rose-au-Rue	1924–30, 1934–36	Husvik
British-Norwegian Whaling	Grady	1927–28, 1930, 1934	*Deane*

Company	*Stations*	*Operating Years*	*Catchers*
Polar Whaling	Hawke Harbour Rose-au-Rue	1937, 1939–40, 1943–51 1937	*Shila, Signa, Silva, Sluga Spina, Morelos, Charcot, Gun VI*
Marine Oils	Rose-au-Rue	1941–49	*Morelos*
Olsen Whaling and Sealing	Williamsport	1945–51	*Finback Morelos Olaf Olsen* (*Rio Sama*), *Gun VI*
Arctic Fishery Products/ Newfoundland Fur Farmers Feed Cooperative	Dildo	1948–68	*Jigger, Dorothy Winter, Arctic Skipper, Arctic Venture, Matthew II, Eriksen, Shirly and Gladys*
Hawke Harbour Whaling	Hawke Harbour	1956–59	*Sposa, Southern Foam*
East Coast Whaling	Dildo	1966–72	*Kyo Maru 17, R.D. Evans, Westwhale 8, Happy Adventure*
Atlantic Whaling	Williamsport	1967–72	*Fumi Maru 15*

Note: Catchers are not station/date specific

APPENDIX 3

Applications for Whaling Licences, Newfoundland and Labrador, 1902–03

Station Location	*Company*	*Date*
Little Burin Bay	CSW	22/07/02
Northwest Arm, Trinity Bay	CMC	02/09/02
Cat Cove, Indian Bay	CMC	09/09/02
Safe Harbour	CMC	28/10/02
Balaena*	CSW	06/12/02
Snook's Arm*	CSW	09/12/02
Bay Chaleur*	NSW	15/12/02
Rose-au-Rue*	NSW	15/12/02
Beaver Cove	HJE	15/12/02
Aquaforte*	AE	22/12/02
Temple Bay, Labrador	BBL	13/01/03
Curlew Harbour, Labrador	UWC	23/01/03
Lark Harbour	CC	26/01/03
Cape Ray	CC	26/01/03
Mark's Harbour, Labrador	CAD	29/01/03
St Barbe's Bay	JRK	?/02/03
St Mary's (Riverhead)*	MC	?/02/03
Hawke Harbour, Labrador	DAR	06/02/03
Port Saunders	JRK	?/02/03
Cape Charles*	BBL	17/02/03
Cape Broyle*	CBT	23/03/03
Quirpon	CC	03/03/03
St Lunaire Bay	NSW	03/03/03
Northwest Arm (Lapoile)	KRP	16/03/03
Curlew Harbour	TNW	23/03/03
Harbour Grace	WHW	23/03/03
Little St Lawrence*	SLW	31/03/03
St Paul's Bay (Dublin Cove)	JB	05/04/03
Bell Island (Grey Islands)	CC	07/04/03
Gold Cove, White Bay	CC	23/05/03
Adlavik, Labrador	BTB	21/07/03
Hopedale, Labrador	BTB	21/07/03
Tuaving Ayualvk, Labrador	CC	21/07/03
Caplin Bay	CC	21/07/03
Okak Mission, Labrador	HDC	21/07/03
Hebron, Labrador	HDC	21/07/03
Raman, Labrador	HDC	21/07/03
Eclipse Harbour, Labrador	HDC	21/07/03
Old Perlican	WBG	04/08/03
Barrow Harbour, Bonavista Bay	WBG	04/08/03
Hooping Harbour, Western Bay	AE	11/09/03

Station Location	*Company*	*Date*
Rencontre, Fortune Bay	EWR	28/09/03
Cottles Cove, Notre Dame Bay	EWR	28/09/03
Portugal Cove	JR	27/11/03
Bell Island, Conception Bay	JR	27/11/03
Blanc Sablon, Labrador (L'Anse au-Loup)*	JB	10/03/03

Source: Newfoundland, *Annual [Fisheries] Report for 1903*,1904

*Denotes whaling stations either active or under construction in 1903. The following abbreviations are used in this table: AE (Anders Ellefsen); BBL (Bowring Bros.); BTB (B.T. Byers); CAD (C. and A. Dawe); CBT (Cape Broyle Trading and Whaling Co.); CC (Colin Campbell); CMC (Colonial Manufacturing Co.); CSW (Cabot Steam Whaling Co.); DAR (Daniel A. Ryan); EWR (E.W. Roberts); HDC (H.D. Crowe); HJE (Henry J. Earle); JB (Job Bros.); JR (Joseph Roper); JRK (James R. Knight); KRP (K.R. Prowse); MC (M. Cashin); NSW (Newfoundland Steam Whaling Co.); SLW (St Lawrence Whaling Co.); TNW (Terra Nova Whaling Co.); UWC (United Whaling Co.); WBG (W.B. Grieve); WHW (W.H. Whitely).

APPENDIX 4

Newfoundland and Labrador Whaling Industry, 1898–1911, Fate of Catchers

Vessel	*Owners*	*Dates*	*History*
Cabot	Cabot Steam Whaling Co.	1898–1914	Local freight, 1915. Scrapped, 1951
Puma	Newfoundland Steam Whaling Co./ Rose-au-Rue Whaling Co.	1904–15	UK war effort, and "Italian Government," 1916
Lynx	Newfoundland Steam Whaling Co./ Rose-au Rue Whaling Co.	1901–15	UK war effort, 1916(?)
Humber	Ellefsen	1902–06	Japan, 1906
*Viking**	Cabot Steam Whaling Co.	1902–03	To Cape Broyle Wh. and Tr. Co., 1904
*Falcon**	Cape Broyle Whaling and Trading Co.	1904–05	Ex-*Viking*, Cabot Steam. Name changed 1904. Japan, 1907
St Lawrence	Dr L. Rissmüller, Little St Lawrence	1903–06	To British Columbia, 1906
*Mic Mac***	Mic Mac Whaling Co./ Phoenix Whaling Co.	1904–06	To Dr L. Rissmüller, Little St Lawrence, 1907
*Sebastian***	Dr L. Rissmüller, Little St Lawrence	1907–08	Ex-*Mic Mac*. Local freight, 1908. To British Columbia Whaling, 1909–16. Towing
Avalon	St Mary's Whaling and Trading Co.	1904–06	To Japan, 1907
Neptune	Neptune Whaling Co.	1905	Fisheries protection. To Japan, 1907
Fin	Atlantic Whaling and Trading Co.	1904–06	To Japan, 1906. Wrecked intransit, Red Sea, 1907
Hump	Atlantic Whaling and Trading Co.	1904–14	Mail service and fishery protection, 1915. Sunk in collision with *Argyle* off St Lawrence, 1915
Hawk	Cape Broyle Whaling and Trading Co.	1904–12	Local freight, 1913. To West Coast U.S.A., 1919
Harbour Grace	Harbour Grace Whaling and Trading Co.	1904	Lost in transit to Newfoundland, 1904

Vessel	*Owners*	*Dates*	*History*
Baccalieu	Harbour Grace Whaling and Trading Co.	1905–06	Replaced Harbour Grace. To Japan, 1907
Port Saunders	Hawke's Bay Whaling Co.	1905–13	Local mail, 1915. To California, 1919
Cachalot (1)	Labrador Whaling and Trading Co.	1905–10	Wrecked, Belle Isle, 1910
Cachalot (2)	Labrador Whaling and Trading Co.	1911–16, 1919	Local mail, 1916. Hawke Harbour, 1919
*Othar****	Ellefsen	1907	Customs service, south coast, 1908. Scrapped at Harbour Grace, December 1932
*Lloydsen****	A/S Cosmos	1912	For combined Newfoundland/East Greenland sealing/whaling

*Same vessel; **same vessel; ***not built specifically for Newfoundland owners

APPENDIX 5

Wages and Occupations of Newfoundlanders Employed at the Aquaforte Whaling Station, 1902–03

Name	*Occupation*	*Months*	*Total Wage ($) (Including Overtime)*
Henry Ryan	flenser	10	339.48
Patrick Byrne	flenser	10	290.03
Peter Croft	flenser	10	313.15
Bert Croft	flenser	5	328.05
Michael Meagher	flenser	11	288.73
James Meagher	flenser	6	123.93
Richard Meagher	flenser	11	319.32
William Meagher	flenser	11	235.98
James Meagher	flenser	11	312.65
William Croft	flenser	6	123.88
Patrick Saunders	flenser	11	224.23
Peter Case	flenser	11	291.72
Henry Winsor	flenser	11	255.68
William Croft	flenser	11	225.24
Frederick Ryan	flenser	11	225.14
John Case	flenser	11	221.89
William Keough	flenser	6	127.84
John Healy	flenser	11	279.21
George Johns	flenser	6	170.16
William Canning	flenser	6	121.52
George Croft	flenser	6	147.11
John Ryan	flenser	11	292.15
James Winsor	flenser	6	151.27
Alan Graham	engineer	11	262.17
James Hiscock	flenser	6	124.18
Patrick Welsh	flenser	6	125.19
Bernard Dale	flenser	11	501.97
Frederick Dale	flenser	11	229.15
John Carter	cooper	6	372.25
William Grimes	blacksmith	4	82.93
Bessie Meagher	housekeeper (?)	–	56.00
Annie Croft	cook	–	64.00
William Croft, Jr.	flenser	5	130.60
John Ryan	flenser	10	247.45
Total Wages:			$7,604.25

Source: Employment and Wage Ledger, 1902–03, Ellefsen papers

APPENDIX 6

Wages and Occupations of Norwegians Employed at the Aquaforte Whaling Station, 1902–03

Name	*Occupation*	*Total wages $ (@ 3.70 Kr.)*
Ole Kristiansen	1st mate	359.73
Oscar Hansen	2nd mate	69.91
Anders Simmerlund	–	140.54
N. Nilsen	seaman	180.00
S. Nilsen	seaman	182.84
H. Gulliksen	seaman	190.27
Anders Andreassen	skipper	180.00
Richard Lorentsen	skipper	190.27
Henrik Ellefsen	skipper	660.61
Johann Andreassen	skipper	343.37
M. Millelsen	skipper	596.50
Kristian Hansen	skipper	1626.22
Carl Larsen	blacksmith	104.05
Anton Sørensen	1st mate	106.86

Source: Employment and Wage Ledger, 1902–03, Ellefsen Papers

APPENDIX 7

Aquaforte Whaling Station: Buildings, Machinery, and Equipment, 1908

Structure	*Items*	
Manager's dwelling house	1 no. 2 Champion stove	1 no. 7 Ideal stove
2 storey wooden, 38 x 28 ft	2 no. 1 Champion stoves	1 iron safe
	2 no. 13 Peri stoves	1 office table
Bunk house – 1 storey wooden,	1 kettle	2 no. 2 Tortoise stoves
38 x 27 ft	2 kitchen chairs	1 no. 6 Ideal stove
	2 boilers	1 stove
	1 coffee kettle	1 large caboose
	2 large frying pans	1 mincing machine
	1 set tea scales	
Factory – 2 storey wooden,	4 boiling tanks, 8x16 ft.	2 large galvanized blocks
54 x 50 ft	4 cooling tanks, 8x8x10 ft.	1 anchor, wooden stock
	4 blubber tanks, about 9x7x7 ft.	1 hand saw
	1 cooling tank, about 16x8x4 ft.	1 ice saw
	1 wooden cooling tank, 12x6x5 ft.	1 cross cut saw
	1 elevator, about 90 ft. long	2 tackles
	shafting and pulleys	1 white painted boat
	2 grindstones	1 boat, flat bottom
	1 set pulley block	5 pr. oars
	1 trolley screw jack	1 steam engine
	1 double winch	1 slicing machine
	1 portable forge	10 wooden buckets
	2 pipe wrenches	2 zinc buckets
	2 pipe cutters	3 funnels
	1 no. 3 1/2 Jerrick dies	2 oil cans
	1 bell	4 flensing knives
Forge, wooden building	1 bellows	1 brace
	1 anvil	1 box stocks and dies
	1 sledge	1 vise
	1 hammer	1 hand vise
	1 ratchet drill	drills, chisels and other
		tools used for
		straightening harpoons
Boiler house, 1 storey wooden,	1 steam boiler	1 steam tube cleaner
40 x 33 ft	1 steam pump	1 round tank,
		about 5 x 6 ft
Coal shed, 61 x 30 ft	4 hatchets	1 bung bore
Store, 61 x 16 ft	10 coal shovels	2 deck brooms
(wooden building)	4 long-handled shovels	1 white headlight
	1 sledge	1 leather hose
	2 pick axes	10 wheel barrows
	4 prongs	

Structure	*Items*
Slip for steamer and belongings	2 heavy heaving-up tackles 2 small blocks 11 big shackles
Slip for whales and belongings, 110 x 60 ft	1 heaving-up winch 1 flensing-up winch 1 canting winch 1 heaving-up tackle 2 heaving-up chains 6 snatch blocks
Stable, wooden, 30 x 22 ft	
Poultry house, wooden, 14 x 19 ft	
Powder magazine, brick building	
Storehouse, wooden, 12 x 14 ft	
Pig house, wooden, 11 x 24 ft	
Cellar, with wooden store on top, 16 x 20 ft	

Source: Elletsen, Anders, to Messrs. F.W. Bissett and Co., 27 June 1908, Ellefsen papers
Note: Furniture and utensils in the manager's house not included. For a list, see Dickinson and Sanger 2001, 109–11.

APPENDIX 8

Portion of Rissmüller's Patent for Extracting Fatty Substances from Meat

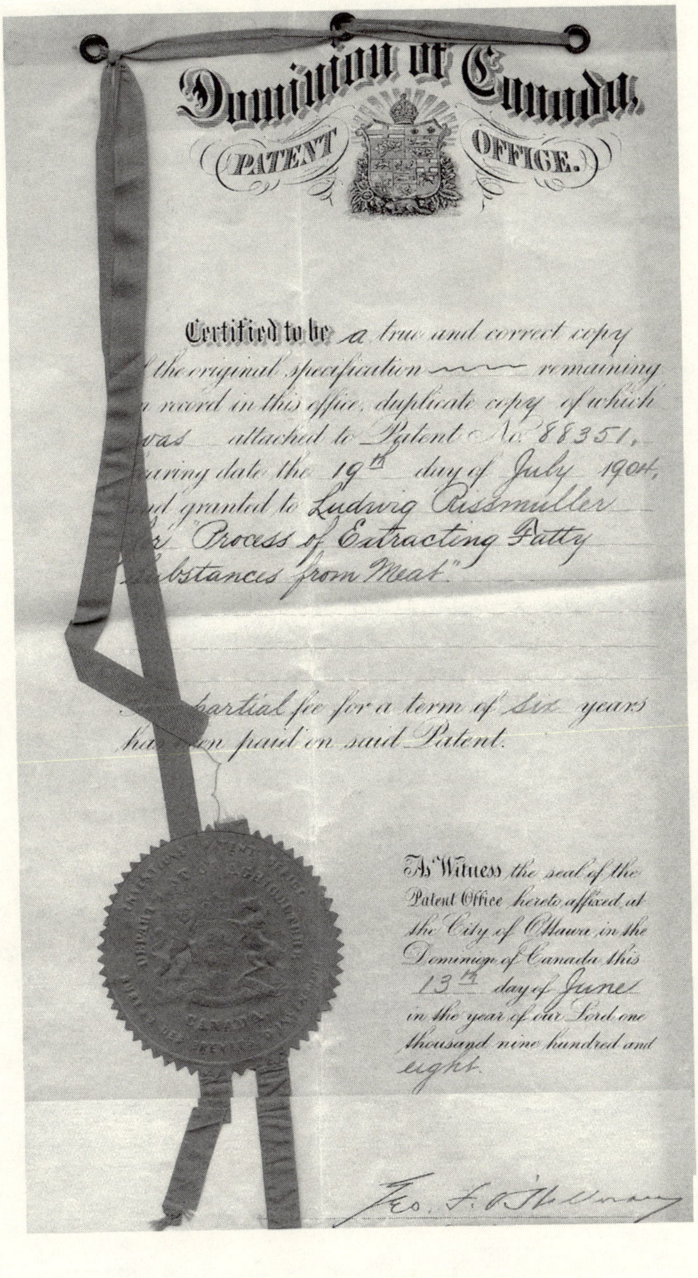
Dominion of Canada

Patent Office.

Certified to be a true and correct copy
[o]f the original specification remaining
[o]n record in this office, duplicate copy of which
was attached to Patent No. 88351,
[b]earing date the 19th day of July 1904,
and granted to Ludwig Rissmuller
[fo]r "Process of Extracting Fatty
Substances from Meat"

[illegible] partial fee for a term of Six years
ha[s be]en paid on said Patent.

In Witness the seal of the Patent Office hereto affixed at the City of Ottawa in the Dominion of Canada this 13th day of June in the year of our Lord one thousand nine hundred and eight.

Geo. F. O'Halloran

COPY.

SPECIFICATION.

TO ALL WHOM IT MAY CONCERN:-

Be it known that I, LUDWIG RISSMULLER, Engineer, residing in the Borough of Brooklyn, County of Kings, City and State of New York, have invented certain new and useful Improvements in

PROCESS OF EXTRACTING FATTY SUBSTANCES FROM MEAT

of which the following is a specification:-

If meat is boiled with water there is produced a coagulation of the albuminous matter, which is in a soluble form in the original cold meat. The greatest part of the coagulated albuminous substances remains as an insoluble body in the meat and closes up the pores hereby obstructing the free circulation of the heated solution through the meat and preventing a thorough separation of the fatty substances from the meat.

-1-

It has been attempted to add acid to the raw meat in the beginning and to thus assist the separation of the fatty matter, but this procedure has great disadvantages. To obtain an effective action of the acid at all, a comparatively large quantity of acid is necessary. If the acid is too dilute it works too slowly and ruins the color of the fatty substances owing to the length of the boiling It is well known that the longer fatty substances or oils are boiled, the darker the color will be. If the acid is too concentrated it darkens the fat very quickly and besides destroys the meat. By my process I prevent the coagulation of the albuminous substances in the meat, liberate the fatty substances quickly and by shortening the time of boiling am able to so extract the fatty substances or oils that they will be of a lighter color. The value of the product is hereby improved as the fats of lighter color are always of higher value than the darker products.

My invention consists of the following process:

To prevent the formation of coagulated albuminous substances I add to the water in which the meat is heated, substances of alkaline reaction, such as caustic soda or caustic potash or carbonate of soda or carbonate of potash. By these chemicals the formation of coagulated albuminous matters is prevented or, if they should have been formed, they will be dissolved again. By the action of the alkaline solution the albuminous substances are extracted from the meat and the solution has therefore a free circulation through the fibrous material of the meat.

The liberation of the fatty substances in hereby hastened and a part of the fatty substances will soon rise to the surface. A part however will not come to the surface by this treatment and can therefore not be separated, as by the action of the caustic alkalies or the carbonates of alkalies a partial saponifying or emulsifying process takes place. It is well known, that the commercial carbonates of the alkalies are not free of caustic alkali and therefore are able to saponify or at least to emulsify a part of the fatty substances. Therefore the soaps or emulsions formed by the alkali treatment must be destroyed, before the total amount of the fatty substances can rise to the surface and be separated. To obtain this result I add, after the alkali has acted, a sufficient quantity of mineral acid, preferably sulphuric acid,. Hereby the alkaline substances are neutralized, the soaps and emulsions destroyed and practically the whole amount of the fatty substances is liberated and forced to the surface, where they can be collected in any suitable manner.

The preferred way I execute my process is the following:-

For the treatment of meat I employ tanks about 12 feet long, 6 feet broad and 4 feet high. They must be constructed of a material which stands the action of the acid employed. The tanks I have in use at present are either constructed of wood alone or of wood lined with lead.

I charge each tank with about one ton (2000 lbs.) of water to which I add about five pounds of caustic alkali or about ten pounds of carbonate of soda or of potash. The tanks are provided with preferably imperforate heating coils made of lead and by admitting steam into these coils the solution in the tanks is heated. At the same time I charge the tank with about five tons (10,000 lbs.) of meat **for the** extraction of the fatty substances. As a rule I raise the temperature in the tanks to the boiling point, but I have found in some cases a lower temperature sufficient for this alkali treatment- 90° centigrade has given good results in some cases. The albuminous substances cannot coagulate in the presence of the hot alkaline solution, but are kept in solution and are extracted from the meat, and the fibrous part of the meat is loosened up by this process, so that the hot solution can permeate it freely and separate the fatty substances of the meat, whose pores are now open and are not clogged up by coagulated albuminous matter. At the same time some part of the fatty matter will be saponified or emulsified and therefore cannot rise to the surface. I will mention here, that the proportions of alkaline substances, water and meat may be altered in order to suit the quality of the meat and I do not confine myself to the exact proportions set forth above. I have found, however, that the proportions mentioned, although the amount of alkali seems to be exceedingly small, have been quite sufficient to produce the desired results. After the boiling or heating of the above alkalie solution has been continued for about three hours, the acid is added, to neutralize the alkali and destroy the soaps and emulsions. The duration

NOTES

ABBREVIATIONS

AMNH	American Museum of Natural History
BCPA	British Columbia Provincial Archives
CITES	*Convention on International Trade in Endangered Species of Wild Fauna and Flora*
COSEWIC	Committee on the Status of Endangered Wildlife in Canada
CNS	Centre for Newfoundland Studies, MUN
DA	*Dundee Advertiser*, Dundee, Scotland
DN	*Daily News*, St John's, Nfld.
EC	*Evening Chronicle*, St John's, Nfld.
EH	*Evening Herald*, St John's, Nfld.
EM	*Evening Mercury*, St John's, Nfld.
ENL	*Encyclopedia of Newfoundland and Labrador*
EP	Ellefsen Papers, Chr. Christensens Hvalfangstmuseum, Sandefjord, Norway
ET	*Evening Telegram*, St John's, Nfld.
GM	*Globe and Mail*, Toronto, Ont.
HAC	*Historical Atlas of Canada*
ICRW	International Convention for the Regulation of Whaling
INWRD	International Network for Whaling Research Digest
IWC	International Whaling Commission
IWS	International Whaling Statistics
JNHA	Journal of the Newfoundland House of Assembly
JP	Job Papers, NSA, MUN
MHA	Member of the House of Assembly, Newfoundland
MMPA	*Marine Mammal Protection Act*, Congress of the United States
MUN	Memorial University of Newfoundland
NAMMCO	North Atlantic Marine Mammal Commission

NFMS	U.S. National Marine Fisheries Service
NLHWS	Newfoundland and Labrador Heritage Web Site, http://www.heritage.nf.ca
NSA	Newfoundland Studies Archive, MUN
NYFG	*New York Fishing Gazette*, N.Y.
PANL	Provincial Archives of Newfoundland and Labrador
PF	*Pacific Fisherman*, Portland, Oregon
PH	*Peterhead Sentinel*, Peterhead, Scotland
PLO/CO	Public Record Office/Colonial Office, London
RG	*Royal Gazette*, St John's, Nfld.
SP	Salvesen Papers, University of Edinburgh Library
TC	*Compass*, Conception Bay Centre, Nfld.
TMC	*Morning Chronicle*, St John's, Nfld.
TT	*Telegram* (formerly *Evening Telegram*), St John's, Nfld.
UBC	University of British Columbia
UFAW	United Fishermen and Allied Workers
VC	*Victoria Colonist*, Victoria, B.C.
VMW	Vancouver Maritime Museum
WS	*Western Star*, Corner Brook, Nfld.

CHAPTER ONE

1 Newfoundland entered the Canadian federation in 1949 as the tenth province. It was renamed Newfoundland and Labrador on 21 October 2003.

2 General overviews of the pre-European occupation of Newfoundland and Labrador are provided in "Aboriginal Peoples" n.d.; McGhee 1996; O'Flaherty 1999, 1–9; "Prehistory" 1987; Renouf 1999; Tuck 1976a; Wright 1995. For more specific information on different indigenous groups, see Maritime Archaic Indians: Tuck 1976b; 1991, 462–5; and 1998; Wright et al. 1987, pl. 6; for Palaeo-Eskimos: "Palaeo-Eskimo Tradition" 1993; Pastore 1998a; Renouf 1993, 185–212; 1994, 145–64; Robbins 1986, 119–24; Wright 1987a, pl. 7; 1987b, pl. 8; for Inuit: Kennedy 1991, 57–65; Pastore 1994; 1998b; Wright 1987c, pl. 9; for recent Indians: Howley 1915; Marshall 1996; Pastore 1992; 1993, 260–77; Renouf 1999, 50–2.

3 Near L'Anse aux Meadows. Macpherson and Wallace 1987, pl. 16, argue that the location and the lack of other confirmed Norse settlements in North America make it "a candidate for identification as Leifsbudir, Leif Eiriksson's settlement or *landnam* in the Vinland of the *Greenlander's Saga*."

4 O'Flaherty 1999, 7–9, provides a concise overview of Norse expansion westwards to Iceland, Greenland, and North America. Macpherson and Wallace 1987, pl. 16, show the scale and scope of Norse migration. For more detailed information including assessments of conflicting evidence, see S.C. Brown 1993, 100–4; Bruun 1918; Gad 1971; A.S. Ingstad 1997; A.S. Ingstad and H.M. Ingstad 1986; H.M. Ingstad 1969; G. Jones 1964; Kristjansson 1970; Macpherson 1997; Magnusson and Palsson 1980; Sanger 1985, 50–5; "Norse in the North Atlantic," 1997.

5 At the beginning of the last millennium, for example, Thorfinn Karlsefni, having sailed west to North America from Greenland and then southward along the coast, wintered at a site probably near L'Anse aux Meadows. This was difficult for the poorly prepared expedition. In the spring they visited a nearby island to hunt and fish, hoping also that it might provide food from "stranded whales" (Magnusson and Palsson 1980, 96).

Shortly thereafter "a whale came in," and although "Karlsefni had a wide knowledge and experience of whales … [he] did not recognize this one" (G. Jones 1964, 180).

6 Europeans may have visited northeastern North America before John Cabot's voyage of 1497, most probably from Bristol, the Basque country, and Portugal. Other possible visitors include an Irish monk, St Brendan (sixth century), a Welsh prince, Madoc (twelfth century), and Henry Sinclair, a Scot from the Orkney Islands (fourteenth century). See, for example, "Before Cabot" 1997; Carus-Wilson 1962; "English Voyages before Cabot" 1997; "Irish Monks and the Voyages of St. Brendan" 1997; McGhee and Tuck 1977; Madden 1981, 788–92; "Other Claims for Pre-Cabot Discovery" 1997; "Atlantic Realm" 1987, 47–51; I. Wilson 1991.

7 Pope 1997a provides a comprehensive assessment of the disputed landfall. See also O'Dea 1988; Williams 1996.

8 For full details of Cabot's 1497 voyage of discovery and subsequent European exploration, see Biggar 1911; Boorstin 1983; Brebner 1933; Crone 1969; Cumming et al. 1971; Harrisse 1892 and 1896; P. McGrath 1978; Morison 1971; O'Flaherty 1999; Quinn 1974, 1977, and 1993; Ruggles 1987.

9 In the early 1600s the English were the only serious rivals to the French-dominated Newfoundland migratory fishery. English-French fishery conflicts that were not resolved until 1904 are discussed in "Atlantic Realm" 1987, 47–51; Brière 1997; Budden 1988; Handcock 1994; O'Flaherty 1999, 58–61; Mannion and Handcock 1987, pl. 23; Proulx 1979; Prowse 1895, 539–55; Turgeon 1981, 295–322.

10 Throughout the transition, Newfoundland remained, as O'Flaherty 1999, 36, points out, "overwhelmingly, tied to the sea, a producer of fish." Despite numerous attempts to "industrialize" after confederation in 1949, the inshore fishery continued to be the largest employer. More than twenty thousand licences were issued in 1982. The industry changed dramatically during the second half of the twentieth century as production of shore-cured salt fish gave way to production of frozen fish (Cuff et al. 1988, 159–65).

11 Federal retraining programs allowed some people to remain in their home communities. Cod stocks did not rebound, and in the summer of 1998 the numbers of fishing licences and fish plants were drastically reduced. Workers were offered inducements to retire or relocate to larger centres, and many communities were left without an economic base ("Fisheries" 1998). In September 2003, only stocks along the south coast of the island supported a small commercial cod fishery (J. Perry, Dept. of Fisheries and Oceans, St John's, pers. com. 2003).

12 "Fisheries" 1998.

13 As L. Harris 1997, 10, indicates, each settlement was located at the edge of the sea where, "clinging like limpets to the rock, the settlers established their holdfasts resolutely turning their backs to the land and their faces to the seas and to the fields of their endeavours that lay submerged before them." The cod fishery not only dictated the pattern of settlement but the catching, salting, drying, and marketing of fish laid down the forms and structures of the society that emerged (ibid.). While early settlements in Newfoundland varied in size and character, most were situated "out-the-bay" where cod could be reached by shallop within a few hours of sailing or rowing (Handcock 1994, 136).

14 Formal settlement experiments by the English began in 1610 when the London and Bristol Company founded Cuper's Cove (Cupids). All failed. See Cell 1969, 1982a, and 1982b, 97–114; Hiller 1997; O'Flaherty 1979 and 1999, 21–37; Prowse 1895, 86–169; Quinn 1982, 9–30. By the 1670s there were thirty different settlements with at least one family in each ("Atlantic Realm" 1987, 48–9). Handcock (1994, 136–7) estimates

that British immigrants created 200–250 settlements up to the 1830s. The 1857 Abstract Census and Return recorded 615 communities, and by 1891 (Census 1893) the number had nearly doubled. The report of 1,447 settlements in 1911 (Census 1914) was the largest number ever recorded. With the decline of the cod fishery, figures fell to about 800 in 1994.

15 As one source ("Atlantic Realm" 1987, 47) reports, "the European fishery had reached rocky, coniferous land with almost no agricultural potential." The harsh reality of an implacably cruel environment in which bare survival was a triumph for early immigrants is best described by L. Harris 1997, 9: "True, the near shore resources of the forest were exploited to furnish the materials for the construction of boats and the basic fishing room facilities of wharves, stages, stores, and flakes, together with modest dwellings. True, small patches were cleared and planted to such hardy crops as potatoes and cabbages, and to hay for the sustenance of a few domestic animals. But as important as were those uses of the land in assisting settlers to eke out a bare subsistence, there was no perception that the land itself could offer a livelihood alternative to the fisheries, or generate a class of land owning farmers and associated artisans who might, as in New England, for example, bring about diversification of employment, raise land values and provide for the accumulation of capital for development."

16 For detailed information on the migratory and residential salt cod fisheries, see Anspach 1819; Cell 1969; Cuff et al. 1988; R. Davis 1973; "Fisheries" 1998; Gilchrist 1972; Handcock 1977, 1981 and 1994; Head 1976; Head et al. 1987; Innis 1954; Lounsbury 1934; Mannion 1982; Mannion and Barkham 1987; Mannion and Handcock 1987; Mannion and Head 1987; Mannion et al. 1987; Marcus 1980; K. Matthews 1968; Miller-Pitt and Russell 1988; O'Flaherty 1999; Ommer 1983 and 1987; "Organizing the Migratory Fishery" 1997; Pope 1997b; Prowse 1895; Quinn 1979a and 1979b; Ryan 1983a and 1983b; Steel 1986; "Atlantic Realm" 1987; "English Fishery and Trade in the 18th Century" 1997; "English Migratory Fishery and Trade in the 17th Century" 1997; "International Fishery of the 16th Century" 1997; Whiteley 1972; Yonge 1963. Janzen 2003 provides a comprehensive review of sources dealing with the Newfoundland cod fishery.

17 The most detailed analysis of the original destinations of English migrants to Newfoundland and Labrador is by Handcock. See, for example, 1977, 1979, 1981, 1989, and 1994. Handcock's findings are presented geographically in HAC: pls. 23 and 25. See also Cell 1969 and 1982a; Janzen 2001a, 2001b, and 2003; Matthews 1978; P. Neary and O'Flaherty 1974; O'Flaherty 1999; O'Neill 1975 and 1976; Prowse 1895; Sanger 1977; "Atlantic Realm" 1987; "Effects of War on Early Settlement" n.d.; "Voluntary Settlement: The Peopling of Newfoundland to 1820" n.d.

18 Irish settlement in Newfoundland has been most thoroughly investigated by Mannion. See, for example, 1974, 1976, 1982, 1986, 1987b, 1988, 1989, and 2001. Other studies include Burke 1971; Corish 1987; Murphy and Byrne 1987; Nemec 1972 and 1991; Prowse 1895. Geographical illustrations of Irish-Newfoundland connections are provided in *HAC* 1987, pl. 25.

19 The operations of the Lester family in Trinity during the eighteenth century were typical of a system that could be run both by agents of English firms or by enterprises based in Newfoundland. The company's trading connections (serving migratory fishermen and the growing resident population) are detailed in Handcock and MacPherson 1987, pl. 26.

20 Under the "truck system," Newfoundland merchants outfitted fishermen for the season on credit in exchange for fish. Many families rarely saw cash. A significant proportion of "income" was produced "in kind" – the production of food, clothing, shelter, and fishing equipment ("Economy" n.d.; Miller-Pitt 1981, 668). There is not full agreement

on the role of the truck system in the peopling of Newfoundland. See, for example, Bates 1994, 428; Cadigan 1995; Crawley 1989, 311–36; Hiller 1990, 90–108; Ommer 1990 and 1991; Pierson 1992, 202–20.

21 For greater details see, for example, Cuff et al. 1988; Handcock 1994; Janzen 2001b; "Decline of the Migratory Fishery" n.d.; Ryan 1983b.

22 Mannion 1987b, pl. 27, shows the growth of St John's and its role in the transformation of mercantile control of the Newfoundland fishery from England to Newfoundland.

23 The resident population increased from 21,975 in 1805 to 40,568 in 1815 (Cuff et al. 1988, 144). The slow pace of Newfoundland settlement in comparison to the general expansion of the British Empire during the seventeenth and eighteenth centuries was due primarily to Newfoundland being considered an industry rather than a colony – as an activity rather than as a society ("Voluntary Settlement: The Peopling of Newfoundland to 1820" n.d.). As O'Flaherty 1999, 33, explains: "The Newfoundland inhabitants were left with no foundation to build on, supposing they wanted to start building an authentic society." It was not until the nineteenth century, consequently, that growing settlements began to attract tradesmen, artisans, professionals, and officials and that legal, political, and social institutions emerged. For information on the late evolution of Newfoundland society, see Baker 1988 and 1993; Bonnycastle 1842; Cadigan 1995; Carson 1813; English 1990; Fay 1956; Field 1924; Fitzgerald 1997; Gosling 1910; Gosse 1890; J.P. Greene 1970; Gunn 1977; L. Harris 1959; Harvey 1899 and 1902; Hatton and Harvey 1883; Hiller 1980; Hiller and Neary 1980; Jukes 1842; Lounsbury 1934; McCann 1988; P. McGrath 1911; McLintock 1941; K. Matthews 1968, 1975, and 1978; Mayo 1948; Morris 1827 and 1848; Moyles 1975; Murray 1968; P. Neary and O'Flaherty 1974; O'Flaherty 1979 and 1999; Pedley 1863; Reeves 1967; Rowe 1964; Ryan 1983b; Seitz 1926; Sider 1980; Talbut 1882; "Emergence of a Newfoundland Society" n.d.; Thornton 1979; Tocque 1844 and 1895; Wells 1966; M. White 1987; Wicks 1974; Willson 1897; W. Wilson 1866.

24 Cuff et al. 1988, 144–6; "Fisheries" 1998; Handcock 1994, 136; Miller-Pitt 1981, 667.

25 As Alexander 1980, 19–20, explains, the inshore salt cod industry dominated Newfoundland's economic history in the nineteenth century and continued to be the single most important source of employment and market income well into the twentieth. In the 1880s, 90 per cent of males were engaged in fishery-related work, and 90 per cent of the colony's exports were fishery products, mostly salt cod ("Fisheries" 1998).

26 Ryan 1980, 42.

27 According to Alexander 1980, 22–1, for example, Newfoundland's per capita income during the early 1880s was approximately half that of Canada. See also Alexander 1974 and 1976.

28 The subsistence pothead "drive" was primarily an opportunistic operation organized on a commercial basis in 1946 when Arctic Fishery Products was formed at Dildo, Trinity Bay, to help ease a food shortage on Canadian mink and fox farms. Access to a relatively reliable food source encouraged the growth of a Newfoundland-based fur industry into the 1960s. For greater details, see Dickinson and Sanger 1999a, 41–5.

29 Although more capital intensive, the development of two new branches of the traditional inshore fishery during the Napoleanic Wars helped support the growing residential population. The Labrador cod fishery and offshore seal hunt were interdependent. Entrepreneurs, mostly from Conception Bay, used schooners to conduct a migratory fishery along the north shore of the island (Petit Nord) formerly controlled by the French. When this area was restored to the French in 1814, the vessels were forced to move further north to Labrador. The new venture was similar to the island fishery, with families who fished from shore called "stationers," while those who remained on

the schooners were known as "floaters." Besides expanding the cod fishery, harp sealing each spring provided additional income for fishermen and vessel owners. The construction of the schooners also provided employment in ship-building trades. For more information on the northern or Labrador cod fishery, see Black 1960; Butt 1998; Cuff 1991; Cuff et al. 1988, 144–6; "Fisheries" 1998; Ryan 1971 and 1980, 44–6; Sanger 1973, 14–8 and 1977, 146–50.

30 The Newfoundland harp seal fishery is documented in C. Brown 1972; Cuff et al. 1988, 144; England 1969; W.H. Greene 1933; Harrington 1986; Hiller 2001a, 2001b, and 2001c; Kean 1935; Ryan 1980, 44–5, 1994a and 1994b; Sanger 1973, 1980, 1988; 1998a, 1998b, 1998c, 1998d, and 1998e.

31 In the 1880s employment levels could not be maintained. The male labour force engaged in catching and curing fish, for example, declined from a peak of 60,000 in 1884 to just under 37,000 in 1891 (Alexander 1980, 23).

32 As the traditional economy approached its maximum growth, government began "to search for a development strategy which would reduce the rate of inshore expansion and initiate its relative decline" (ibid., 15). The economic strategies of successive governments have been studied by Alexander (1974, 1976, 1977, 1980, and 1983) and House (1985, 1986, 1988, 1998a, and 1998b). For additional information, see Antler 1975; Bursey 1993; Cuff 2001; Economic Council of Canada 1980; "Economy" n.d.; "Government and Politics" n.d.; Harrington 1994; Hodgson 1981; Iverson and Matthews 1968; F.L. Jackson 1984; Miller-Pitt 1981; Moores 1993 and 1994; P. Neary 1978, 1982, 1988 and 1997; Rennie 1998; J.A. Webb, 2001a, 2001b, and 2001c.

33 Boston was the main destination and principal port of entry for Newfoundlanders seeking employment in the various New England states before confederation.

34 As Blake 1993, 239, explains, life for those who laboured in poverty and deprivation in the isolated outports and bays of Newfoundland in 1949 had changed little since their ancestors arrived. For specific examples, see Grenfell 1895, 1906, 1909, 1910, 1912, 1916, and 1919; D. Jones 2000; Rompkey 1991 and 1996.

35 J.A. Webb (2001a, 2001b) provides a general overview.

36 For a full discussion of this model, see Sanger 1985, 42–6.

37 According to Jenkins (1971, 47), "The successive phases in the history of whaling … have been, for the most part, only possible because either a new species has been attacked, or a new harvest of a previously attacked species has been discovered."

38 The ban was to be implemented in 1986. However, whaling still continues by indigenous peoples and commercially, including in the guise of "scientific whaling." For the current status of whaling in the North Atlantic, see epilogue.

39 There are numerous histories of global whaling. The overview of the traditional and modern components of the industry outlined in this volume is based upon Sanger 1985, 42–182. See also Basberg et al. 1993; Ellis 1980 and 1992; Francis 1990; MacKintosh 1965; L.H. Matthews 1978; Stoett 1997; Tønnessen and Johnsen 1982.

40 For a discussion of the possible origins of European-based commercial whaling, see Sanger 1985, 47–57.

41 Right whales are classified into the family Balaenidae. The IWC (http://www.iwc.org/Cetacea.htm) identifies three species, the North Atlantic right, *Eubalaena glacialis*; the Southern right, *E. australis*; and the North Pacific right, *E. japonica*. Also included is the bowhead whale, *Balaena mysticetus*. This contrasts with, for example, Watson (1981, 68–75), who ascribes to the family the great right whale *Balaena glacialis* (northern and southern hemisphere populations, Atlantic, and Pacific); the pygmy right, *Caperea marginata*; and the bowhead, *B. mysticetus*. Watson also lists other common names for *E. glacialis*, including black right, Nordkaper, and Biscayan right.

42 The Bay of Biscay whale fishery had evolved by the twelfth century into a sophisticated, relatively large-scale, commercial enterprise. See, for example, Agirreazkuenaga Zigorraga et al. 1980; Ciriquiain-Gaiztarro 1961; De Cortazar 1966. Good English treatments of early Basque whaling are provided by Douglas and Bilbao, 1975, 51–7; Harmer 1928; Markham 1881.

43 For a full assessment of evidence supporting this explanation, as well as other interpretations, see Sanger 1985, 59–66.

44 Barkham's analysis of original Basque manuscripts has provided the greatest detail of this previously little known whale fishery: 1974, 1976, 1977a, 1977b, 1978, 1980, and 1982. Her work also led to excavation of the Basque whaling station at Red Bay by Memorial University's Archaeology Unit and Parks Canada. See, for example, Grenier 1985; Tuck 1985; Tuck and Grenier 1981 and 1989.

45 Hakulyt 1965, vol. 8, 10.

46 Purchas 1906, vol. 19, 425.

47 A small number of vessels continued to visit as late as the summer of 1632 (Barkham 1981).

48 Sanger 1985, 69. See also R. Davis 1973; Rowen 1972.

49 Sanger 1985, 72.

50 The Spitsbergen and other phases of Northern whaling have been documented by Bobe 1952; Conway 1900, 1904, and 1906; Dalgard 1962; De Jong 1978; De Veer 1953; Dietz 1923; Edmondston 1809; Edmundson 1911; Egede 1818; Elking 1859; Gerritsz 1904; G. Jackson 1978 and 1981; Laing 1822; Leslie et al. 1831; Lubbock 1937; Martens 1855; Mathieson 1934; Phillips 1819; Powys 1927; Sanger 1985, 72–122, 358–627; 1991a, 1991b, 1991c, 1994a, and 1994b; Scoresby 1820.

51 Many of the basic elements characterizing earlier whale fisheries can also be identified in the origins and evolution of the American whaling trade. The seasonal appearance of large numbers of North Atlantic right and sperm whales close onshore, for example, made strandings a relatively frequent occurrence. These were just as valuable to the early settlers of New England as they had been to the coastal populations of continental Europe, Iceland, and Greenland. Consequently, as G.M. Allen (1928, 340) pointed out, "in these early years dead whales were so often cast upon shore, that laws and regulations for deciding their ownership were constantly being enacted." The colonists quickly developed sophisticated equipment and procedures, however, and "during the third quarter of the seventeenth century, the deliberate pursuit of whales was begun by communities on the eastern end of Long Island, on Martha's Vineyard, on Cape Cod, in the Bermudas, and on Nantucket" (Hohman 1928, 25).

52 The American sperm and right whale fisheries have been extensively studied. The more comprehensive nineteenth century investigations include Clark 1887; Scammon 1874; Starbuck 1878.

53 Relationships between the early American whalers and the colonial administrators in St John's were not always amicable. Governor Hugh Palliser in a letter of 7 August 1776 to Governor Bernard of Massachusetts complained: "The great trouble and difficulty I meet with in keeping good order amongst the fishers in a part of this government [Labrador] is occasioned chiefly by a number of disorderly people from your Province … [enclosed advertisements] to be put up in the Towns within your Government where the vessels employed in the whale fishery mostly belong" (Prowse 1895, 327).

54 Dickinson and Sanger 2001, 3, 16. See also Macy 1835; Sanger 1985; Sanger and Dickinson 1994; Starbuck 1878; Tocque 1895.

55 The dead animal was found floating between Crapaud Point and Cape St Mary's Bank (*EH*, 9 July 1897).

56 *EH*, 19 May 1895.

57 *EH*, 9 September 1897.

58 Tocque 1895, 70. See also Dickinson and Sanger 2001, 16.

59 *WS* (reported in *EH*, 7 April 1903). See also Prowse 1895, 298–9.

60 *EH*, 5 January 1899. The facilities included "Trying-out Houses" on Whale Island at the entrance to Gaultois Harbour.

61 *EH*, 26 February 1895.

62 *EH*, 27 May 1895.

63 The company built their factory in St John's following the end of the 1876 season (*DA*, 7 February 1877). A brief account of their involvement in Northern whaling and Newfoundland sealing is provided in a history printed by the company (Stephen 1932).

64 The company's St John's premises were opened in 1878 (*TMC*, 21 February 1878). The Dundee Seal and Whale Fishing Company deployed two steam-powered ships on the Newfoundland harp seal whelping grounds in 1862 (*DA*, 17 February 1862). Harsh environmental conditions limited their success (*DN*, 2 April 1862). Nevertheless, Newfoundland companies quickly converted their fleets to steam, dramatically altering the local seal fishery (Sanger 1973). The Scots then decided to concentrate their sealing efforts at East Greenland until 1876 when reduced catches again caused them to send a vessel to Newfoundland (*PS*, 19 April 1876).

65 The involvement of Scottish whaling companies in the Newfoundland seal fishery after 1862 is discussed in Sanger 1980 and 1988.

66 Sanger and Dickinson 1989, 131.

67 See, for example, *EH*, 8 June 1897; 18 January, 27 June 1898.

68 L.E. Davis et al. (1997) argue that competition from new sources of illumination was more responsible than stock depletion for the decline in the industry after 1850.

69 The final phases of bowhead whaling in the western Arctic are discussed in Bockstoce 1977 and 1978.

70 As Hohman (1928, 70) points out: "The older whaling fields had been successively worked out and were mostly deserted, but the possibility of capturing Rorquals gave new life to the industry."

71 L.H. Matthews 1968, 189. Logbooks and journals of British northern whalers (see Sanger 1985, 638–42) contain numerous references to sightings of "finners" and the difficulty of killing them. The futility of attempting to catch anything other than a "right" whale is expressed by the surgeon of an Aberdeen whaler who explained in 1831 that "they are seldom or never taken as they are so swift and strong that they would soon run out the lines in a ship" (*Hercules*, 30 April 1831). See as well Gillies 1826, vol. 2: 15; Elking 1859, 79; Sanger 1991c, 128–32.

72 The most comprehensive treatment of the development of modern whaling is provided by Tønnessen and Johnsen, originally published in Norwegian (vols. 1–4, 1959–70). While the English translation (1982) is considerably abridged, it extends information on the industry to 1980.

73 Schmitt et al. (1980) describe Roys's attempt to revitalize traditional whaling. Although Roys and his partner, G.A. Lilliendahl, operated a land-station in Iceland during the 1860s, the equipment and strategies they employed produced generally unfavourable results.

74 Tønnessen and Johnsen (1982, 25–41) provide a detailed description, with photographs, of the whaling system developed by Svend Foyn.

75 Three years later, according to Gaskin (1972, 38), there were "no less than 19 whaling stations on the coast of Norway operating a total of 35 chasers." Niemi (1993) describes the origin and development of modern whaling along the Norwegian Arctic coast.

76 Tønnessen and Johnsen 1982, 66–7.

77 Stations were quickly established around the world. See, for example, Harmer 1928, 69–70; MacKintosh 1965, 145–7.

78 Tønnessen and Johnsen 1982, 95–101, 263–76. For details on the role Spitsbergen played in the introduction of factory vessels, see Arlov 1993.

79 Tønnessen and Johnsen 1982, 367–471.

80 Ibid., 472–686.

81 See, for example, K.R. Allen 1980.

82 Blandford, Nova Scotia, and Dildo and Williamsport, Newfoundland.

83 See, for example, Meredith and Campbell 1988; Waring et al. 2001.

84 See, for example, Mansfield 1985; Waring et al. 2001.

85 See, for example, Mitchell and Reeves 1983; Hay 1985; Waring et al. 2001.

86 Mitchell 1972a and 1975b.

87 Waring et al. 2001.

88 Sanger (1985, 228–32) discusses the importance and changing uses of whale products.

89 J.A. Allen 1908, 278.

90 Ommanney 1971, 61.

91 See, for example, Scoresby 1920, vol. 2, 420–1. As Colwell (1970, 1) points out, whales were "almost the only source of a pure, clean oil which could be used for illumination, instead of sputtering, smelly tallow candles or the costly beeswax candles."

92 See, for example, Sanger 1985, 230.

93 Tønnessen and Johnsen 1982, 227–40. Whale oil was used as a high quality lubricant in the manufacture of cordage, the preparation of rough fibres such as jute, and finer woollen clothes, as an ingredient in paints and varnishes, and as a preservative with tar in the construction of wooden vessels. Whale oil also functioned as a leather softener and was used in the preparation of such a widely divergent product as perfume (Sanger 1985, 230).

94 Tønnessen and Johnsen (1982) provide a good description of oil extraction processes. For a case study of shore-station procedures, see Basberg's discussion of South Georgia whaling between 1925 and 1960 (1993). Wexelsen (1993) provides similar details of practices and working conditions on board floating factories. See also Dickinson and Sanger 2001, 13–15.

95 Factors that contributed to oil quality are discussed in Tønnessen and Johnsen 1982, 38, 48, 99, 142–3, 253–4, 694, 708–12, 719.

96 Whalebone was also used as springs for chairs, sofas, and carriages; in umbrellas, whips, fishing rods, walking canes, travelling trunks, hat boxes, and covers for telescopes; for making brushes (especially chimney sweeps' brooms); as upholstery stuffing; and in a wide range of minor items such as pen holders, shoe horns, and combs (Sanger 1985, 231–2).

CHAPTER TWO

1 *EH*, 15 December 1897.

2 Tønnessen and Johnsen 1982, 83–4.

3 As Innis (1954, 454) explains: "The bad fishery of 1893 and the depression of the nineties had serious results for a community dependant on a single export and heavily burdened with railroad debts and charges." For a full discussion of the state of the Newfoundland fishery at the turn of the century, see "Capitalism in Newfoundland, 1886–1936," ibid., 449–83.

4 A St John's newspaper (*EM*, 20 September 1887) described Norwegian attempts to develop cod fish hatcheries. See also Baker et al. 1992, 26.

5 Tønnessen and Johnsen 1982, 102.

6 Moses Harvey was a member of a prominent St John's merchant family and an eminent clergyman, local historian, and naturalist. For an account of Harvey's life, see Aldrich 1988, 849–50.

7 *Report of the Fisheries Commission Appointed by His Excellency the Governor-in-Council to Investigate the Operations of Fisheries Departments in Other Countries*, 1888.

8 Ibid.

9 *Act to Provide for the Formation of a Fisheries Commission and for Other Purposes*, 1889. *RG*, 2 July 1898. See also *Annual Report of the Newfoundland Fisheries Commission for the Year 1889*, 1890.

10 Nielsen's activities in Newfoundland are described in Baker et al. 1992. The operation of the hatchery, events leading to its construction, and Nielsen's involvement have also been summarized in Harrington 1987a.

11 Newfoundland, *Annual [Fisheries] Report for 1897*, 1898.

12 Cabot Steam Whaling Co. Ltd, 1896–1914.

13 Ibid.

14 *EH*, 28 April 1898.

15 *EH*, 9 December 1897.

16 Ibid.

17 Cabot Steam Whaling Co. Ltd, 1896–1914.

18 Nielsen succumbed to asthma in the Crosbie Hotel where he lived in St John's. He was buried in the General Protestant Cemetery, where his headstone is still visible (Newfoundland, Department of Health, death certificate of Adolf Nielsen, 1903).

19 *EH*, 19 November 1897.

20 *EH*, 13 May 1898. Nielsen appears to have been unpopular amongst a segment of the local population, with the *Evening Telegram* making particularly virulent attacks against him. The rival *Evening Herald* reported that the people "rubbed their eyes with astonishment on Wednesday evening when they saw the 'Telegram' congratulate Mr. Adolph Nielsen on his new whaling venture and devote a column of its space to describing the method of prosecuting the new industry. For years past he has been the target for the incessant abuse of that vile sheet, and its hypocritical professions of friendliness towards him now can only be ascribed to the fact that its masters, the Messrs. Harvey, are large shareholders in the whaling company."

21 *EH*, 19 November 1897.

22 *EH*, 18 January 1898.

23 *EH*, 8 June 1897. The company also operated a tannery, processing seal and porpoise skin into leather (*EH*, 24 August 1895). For details of A. Stephen and Son's operation in St John's, see chapter 1, Note 65.

24 *EH*, 18 January 1898.

25 *EH*, 18 March 1898.

26 *Skimmer of the Waves* was an old barque that had been towed into St John's "many years since" and abandoned. She was said to be haunted and to have been a pirate vessel (*EH*, 11 May 1898).

27 *EH*, 4 and 7 May 1898.

28 *EH*, 2 and 7 June 1898.

29 See, for example, *EH*, 22 April 1898.

30 A letter from Ferryland stated that "a large school of whales has made its appearance there and had been spouting in all directions"(*EH*, 6 May 1898). Whales were also

abundant between Cape Spear and Cape Race (*EH*, 12 May 1898).

31 *EH*, 31 May 1898.

32 *EH*, 10 and 11 June 1898. She was towed by the steamer *Glover.*

33 *Mercantile List and Maritime Directory* 1904.

34 *EH*, 30 May and 17 June 1898. *Cabot* cost $21,000 to build.

35 *EH*, 25 August 1898.

36 *EH*, 21 June 1898.

37 Local newspapers again reported that "enormous schools of whales were lately seen between Cape Spear and Race" (*EH*, 12 May 1898).

38 *EH*, 25 June 1898. At that time Norwegian whaling captains frequently acted as harpooners. For a more detailed discussion of the company's first operating season, see Sanger and Dickinson 1989.

39 *EH*, 27 June 1898. According to Millais (1907, 163), the whale was killed on the previous day.

40 *EH*, 28 June 1898.

41 *EH*, 1 July 1898.

42 *EH*, 28 June 1898.

43 *EH*, 25 and 28 June 1898.

44 *EH*, 29 June 1898.

45 The development of the station is described in *EH*, 25, 28, 29 June, 18 and 26 July 1898.

46 *EH*, 5 and 8 July 1898.

47 *EH*, 8 and 16 July 1898.

48 *EH*, 16 July 1898. Details of the individual voyages are also included.

49 Hot weather at New Perlican, for example, caused seepage from the decaying blubber, and *Cabot* was forced to return to St John's even though she could "hold about 3 fish more" (ibid).

50 Ibid.

51 *EH*, 8 July 1898. Bull saw four whales while steaming to this new location, and the coastal vessel *Portia* encountered "thousands" between Cape Pine and Cape Race on the same day (*EH*, 19 July 1898).

52 *EH*, 25 July 1898.

53 Ibid.

54 *EH*, 30 July 1898.

55 *EH*, 27 July 1898. Despite his desire to catch whales, Bull became involved in a salvage operation when *Cabot* towed *Petunia* off rocks near Aquaforte. Sir W.V. Whiteway, QC, holder of twenty-three shares in the whaling company, obtained $2,500 through Admiralty Court for the master, crew, and owners (*EH*, 26 August 1898).

56 *EH*, 28 July 1898.

57 *EH*, 5 August 1898.

58 *EH*, 6 August 1898. Trinity, Bonavista, Notre Dame, and White Bays are located along the northeast coast of the island (Fig. 1.1).

59 *EH*, 1, 3, 4, 8, 13, and 18 August 1898.

60 *EH*, 20 and 22 August 1898.

61 *EH*, 22 and 23 August 1898.

62 *EH*, 20 August 1898.

63 *EH*, 26 August 1898. The *San Juan* was then used to carry coal from Sydney, N.S., to Snook's Arm (*EH*, 8 October 1898). She was wrecked on 15 August 1900 off Forteau while supplying lighthouses on the Labrador coast (*EH*, 25 August 1900).

64 *EH*, 3 September 1898.

65 *EH*, 19 September 1898.

66 Ibid. Dr F.W. True, head curator of biology at the Smithsonian Institution in Washington, D.C., visited Snook's Arm in late August and early September. He recorded that twenty-five fin and three humpback whales had been killed and that whalers observed more humpbacks after August (True 1904, 111, 212). In another description of the whales killed at Snook's Arm, True included photographs, "the only ones representing living whales in American waters thus far published" (True 1903). For a full discussion of True's interest in Newfoundland and Labrador shore-based whaling, see Sanger and Dickinson 1991a.

67 *EH*, 2 November 1898. True (1899) recorded that fin whales left Green Bay at the end of October, possibly due to the migrations of their main food, cod and capelin.

68 *EH*, 19 November 1898.

69 *EH*, 28 November 1898. The first season's catch is erroneously recorded as forty-seven by Millais (1907).

70 *EH*, 1 October 1898.

71 *EH*, 17 October 1898.

72 *EH*, 1 October 1898. *Damara* sailed three days later (*EH*, 4 October 1898).

73 Newfoundland, *Annual [Fisheries] Report for 1903*, 1904, 8.

74 *EH*, 25 January 1899. See also *EH*, 17 June 1898.

75 *EH*, 14 October 1898. An early visitor to Snook's Arm noted that the local inhabitants "all eat it – up and down the coast they come for it – and they thrive on it" (Duncan 1904, 680).

76 *EH*, 25 October 1898.

77 See, for example, *EH*, 2 and 12 November 1898.

78 Harvey and Co. and the Cabot Steam Whaling Co. subsequently sued for outstanding debts (*EH*, 19 October 1901).

79 *EH*, 14 October 1898.

CHAPTER THREE

1 *EH*, 23 December 1898. For a more detailed discussion of the original period of expansion and consolidation, see Dickinson and Sanger 1990.

2 *EH*, 1 December 1898.

3 *EH*, 4 January 1899.

4 On the recommendation of the postmaster general, the name of Bonne Bay in Hermitage Bay was changed to McCallum Bay and the name of the settlement at the head of the bay to Balaena (Newfoundland, Minutes of the Executive Council, 13 September 1899).

5 *EH*, 4 January 1899. At an average of $80 per ton, it would generate some $50,000 a year (Tønnessen and Johnsen 1982, 753).

6 *EH*, 4 January 1899.

7 *EH*, 23 December 1898.

8 *EH*, 30 December 1898.

9 This may have been the last voyage for *Skimmer of the Waves*. She sank at Balaena, where in 1902 her masts were still "above water" (*EH*, 4 December 1902). During this voyage she carried "two new boilers, three new and a number of old tanks, with 20,000 brick, lime, cement, etc., for the building" (*EH*, 24 March 1899). Construction was to be handled by "James Elliott and J. Smitten, carpenters, of Twillingate, who have worked on the Snook's Arm establishment." The three "new" donkey winches were salvaged from the wrecked *Arbela* (*EH*, 13 April 1899). See also *EH*, 21 January, 21 February,

28 March 1899 for additional construction reports.

10 *EH,* 21 February 1899. *San Juan* was later sent to Rose Blanche to be fitted out for "the gulf seal fishery."

11 *EH,* 24 April 1899.

12 *EH,* 13 June 1899. Angel and Co. had also built and installed the boilers at Snook's Arm the previous year.

13 *EH,* 5 September 1899.

14 *EH,* 28 August 1899. The Balaena factory was completed in late September (*EH,* 18 October 1899).

15 *EH,* 14 February 1899. The weather was "terrible," with "intense cold."

16 Ibid.

17 *EH,* 24 April 1899. See also Southwell 1899. Blue whales were often called "sulphur bottoms" because of the colour produced by micro-organisms growing on the ventral surface.

18 *EH,* 24 April 1899.

19 See, for example, *EH,* 10 April, 3 May 1899.

20 The hull under the waterline had been "badly rubbed by contact with ice" (*EH,* 26 April 1899).

21 *EH,* 10 April, 3 May, 29 June, 22 August, 11 and 22 September 1899.

22 *EH,* 25 June 1899.

23 *EH,* 19 June 1899.

24 For a full account of whaling at Snook's Arm during 1899, see Dickinson and Sanger 1990, 95–101. The season also had a significant impact on the American scientific community and the general public at large (Sanger and Dickinson 1997).

25 *EH,* 28 August 1899.

26 *EH,* 25 October 1899. *Virginia Lake* "spent 12 hours among the putrefying whale carcasses, and all on board say it was an experience they never want to repeat" (*EH,* 7 September 1899).

27 The oil was shipped on *Waldensian* (*EH,* 11 November 1899). Glasgow became important when a Norwegian immigrant to Scotland, Christian Salvesen, established business relations with Foyn and assisted in marketing Norwegian oil through that port. See, for example, Elliott 1998; R.L. Webb 1988a and 1988b. Dundee by this date had become the centre of the declining traditional Scottish whale fisheries at East Greenland and Davis Strait (Sanger 1980, 1985, and 1988).

28 *EH,* 21 October 1899. Millais (1907, 163) reported ninety-five killed. However, Southwell (1899, 108–12) also reported 109 whales. The species composition is variously recorded. Tønnessen and Johnson (1982, 102) estimate one-third fin whales, one-third humpback, and the rest blue and sei whales. Southwell (1899, 108–12) reported eleven blue whales killed from Balaena and nine humpbacks from Snook's Arm, the remainder being fin whales.

29 Oil production began at Balaena in October 1899, where "the factory is finished and the machinery has been running these three weeks past, converting eleven whales captured last winter into oil" (*EH,* 18 October 1899).

30 Based on the six forty-gallon barrels per ton standard used elsewhere in British colonial fisheries. See, for example, Headland 1984; *Report of the Interdepartmental Committee on Research and Development in the Dependencies of the Falkland Islands*, 1920.

31 Cabot Steam Whaling Co. Ltd, 1896–14.

32 Millais (1907, 184) estimated that $60,000 was required for a Newfoundland whaling operation in the early twentieth century.

33 *EH,* 6 May 1902.

34 Sanger and Dickinson 1989, 143–4.
35 *EH*, 28 October 1899.
36 *EH*, 13 November 1899; Southwell 1900.
37 *EH*, 13 November 1899.
38 *EH*, 3 January 1900.
39 *EH*, 30 April 1900.
40 *EH*, 23 May 1900.
41 *EH*, 30 July, 23 October 1900.
42 *EH*, 13 September 1900.
43 *EH*, 6 August, 23 October 1900.
44 Millais (1907, 163) reported a catch of 111, which he seems to have confused with the previous season. True (1904, 156 and 213) recorded sixty-seven blue whales and eleven humpbacks killed from Balaena, the remainder being fins.
45 See, for example, R.L. Webb 1988b, 144.
46 Tønnessen and Johnsen 1982, 102. The success of the 1899–1900 season was marred when Nielsen's health failed and he had to leave St John's to recuperate in Norway (*EH*, 1 February 1900). There were also labour problems at Balaena, with ten men returning to St John's "having struck for higher wages. They were receiving 80 cents and wanted $1 a day" (*EH*, 30 April 1900).
47 The company produced approximately 3 per cent of global whale oil (87,300 barrels) in 1900 (Tønnessen and Johnsen 1982, 228).
48 *EH*, 25 September 1900.
49 A Fortune Bay resident concurred with the selection of the new site, noting in a letter to the *Evening Herald* that there were "plenty of these 'fish' in the southern waters, and two whaling stations some distance apart would do well, but if they were close together the reverse would be the result" (*EH*, 13 September 1900).
50 *EH*, 8 October 1900; 8 January 1901.
51 *EH*, 23 April 1901. They were made by Angel and Co. in St John's.
52 *EH*, 3 June, 11 October 1901. The wooden tank was built by the Horwood Lumber Co. of St John's.
53 *EH*, 5 July, 7 December 1901.
54 Millais 1907, 184. Rissmüller became an important figure in the early twentieth century industry, operating or owning guano production plants at several whaling stations in Newfoundland and British Columbia. His contribution to the industry is described in Sanger and Dickinson 1991a and 1991b; R.L. Webb 1988b, 156–61.
55 *EH*, 26 September 1900; 3 April 1901. *Puma* was to be "a whaling steamer similar to the *Cabot*" (*EH*, 18 January 1901).
56 *EH*, 15 and 25 June 1901. A Norwegian crew brought her to Newfoundland.
57 *EH*, 25 June 1901.
58 *Puma* returned to St John's on 6 October 1901, "having need of repairs. She has done well this summer and hopes to do even better" (*EH*, 7 October 1901).
59 "To have an inspection made to her shaft, and for a general clean up and repairing" (*EH*, 19 November 1900).
60 Only nine had been killed up to the end of April, and a report from Hermitage stated that "it is thought whales will be scarce this season." (*EH*, 1 May 1901).
61 *EH*, 25 June 1901.
62 Ibid. An early July report indicated that *Cabot* was "doing good work in Fortune Bay since the spring opened, and daily kills two and three fish off the harbours along the coast. She has taken to date 50 fish of the fin back and sulphur bottom varieties" (*EH*, 5 July 1901).

63 *EH*, 23 July 1901.

64 *EH*, 29 July 1901.

65 *EH*, 5 August 1901. The strong Harvey connection with this company specifically and Newfoundland whaling in general is shown by the fact that *Cabot* "hauled down to Harvey's premises to take in supplies."

66 *EH*, 1 October 1901.

67 *EH*, 23 November 1901.

68 Mitchell and Reeves (1983, 166) state that all were caught from Reuben's Cove when estimating a season's catch of ten humpback whales for Balaena. True (1904, 156–213) recorded seven humpback and seventy-one blue whales captured at Balaena during this season. The importance of blue whales to company success is indicated by the estimate of Millais (1907, 165) that each yielded more than one hundred barrels of oil worth some £24,000 at the prevailing prices of £20 per ton. See also Tønnessen and Johnsen 1982, 753. Data from the Falkland Islands' industry in 1914–15 give average yields of seventy to eighty barrels for blue whales, thirty-five to fifty for fin, twenty-five to thirty-five for humpbacks and ten to fifteen for sei (*Report of the Interdepartmental Committee on Research and Development in the Dependencies of the Falkland Islands,* 1920, 52).

69 *EH*, 22, 25 January, 25 February 1902.

70 See, for example, *EH*, 26 October 1901; 4 and 18 March 1902.

71 See, for example, *EH*, 31 October, 6 November 1901.

72 *EH*, 22 November 1901.

73 *EH*, 6 November 1901. For additional comment regarding foreign investment in local whaling companies, see two letters, one from a Newfoundlander opposed to Norwegian participation and the other from a Norwegian, Anders Ellefsen, explaining why he wanted to move to the island and build a station at Aquaforte (*EH*, 26 and 31 October 1901).

74 *EH*, 22 November 1901. Bond promised to consider favourably the request for a restriction.

75 *EH*, 21 February 1902.

76 Ibid. Not surprisingly, whaling companies opposed the proposed legislation.

77 See, for example, *EH*, 25 February, 3, 5, 8, 10, 11 and 13 March 1902. The wording of the act was also intended to allay any fears the French might have that the expanding industry would encroach on their "fishing" rights on the Treaty Shore. Governor Cavendish Boyle pointed out in correspondence to the Foreign Office in London that the act would specifically refer to whaling, since the use of "fishery" might "be calculated to attract the attention of the French Government, provoking probably, representations and protests which might otherwise be avoided" (Newfoundland, Minutes of the Executive Council, 24 March 1902).

78 *EH*, 25 February 1902. The industry "should be open to every citizen" (ibid).

79 *EH*, 11 March 1902. Similar concerns had been expressed in Norway, culminating in riots in 1904 directed towards the industry (R.L. Webb 1988b, 139–40).

80 *EH*, 5 and 8 March 1902.

81 *EH*, 10 March 1902. The opposition also wanted legislation on the conduct of the industry to be delayed until the commission reported.

82 *EH*, 5 March 1902.

83 *EH*, 15 March 1902.

84 See, for example, *EH*, 3 April 1902.

85 Guano was valuable as a fertilizer. "Made from flesh and about one-third bone, it contained some 8.5 per cent ammonia and 21 per cent triphosphates," while "the whole of the dried carcass may also be made into one product, forming a rich guano and con-

taining from 10 to 12 per cent ammonia and 17 per cent to 24 per cent phosphates" (Bruce 1907, 483). See also the discussion on Rissmüller's methods in R.L. Webb 1988b, 156–61.

86 See reports carried in *EH*, 24 and 25 April 1902.

87 Newfoundland, *Acts of the General Assembly of Newfoundland: Passed in the 2nd Year of the Reign of His Majesty King Edward VII*, 1902.

88 *Revised Statutes of Canada*. R.L. Webb (1988b, 151) is of the opinion that this amendment was contrary to the Newfoundland act.

89 A British report provides details of the Falkland Islands Ordinance 3, 1906, the amending Ordinance 5 of 1908, and the beginning of the South Georgia industry (*Report of the Interdepartmental Committee on Research and Development in the Dependencies of the Falkland Islands*, 1920).

90 True 1904, 113, 156.

CHAPTER FOUR

1 On 9 April 1902, for example, the *Evening Herald* reported that the schooner *Irene*, recently arrived from Ferryland, had encountered large numbers of whales "all the way down the shore … and if the *Cabot* or *Puma* were about they would have secured good fare." For greater details of the 1902–03 whaling seasons, see Sanger and Dickinson 1991a.

2 *EH*, 14 March 1902.

3 *EH*, 1 and 30 April 1902. The factory was considered "a spacious one and will be rushed to completion."

4 *EH*, 14 March, 16 April 1902. In the interim, equipment needed to build a guano plant was sent to the site and erected under the supervision of Dr Ludwig Rissmüller, later a company shareholder (*EH*, 23 May, 11 September 1902).

5 *EH*, 29 September, 12 November, 5 December 1902. A house also had to be destroyed because the previous occupants had an unidentified infectious disease (*EH*, 13 November 1902). The company may have used its workers at Snook's Arm and Balaena, then operating at capacity (*EH*, 31 May 1902).

6 Some investors held shares in other whaling companies, particularly the Harvey family and the estate of Adolph Nielsen in the Cabot Steam Whaling Co.

7 *EH*, 27 May 1902.

8 *EH*, 22 October 1902. Most were caught in Chaleur and Placentia Bays, including "a whale killed at Lapoile … [that] had an American harpoon in its body with the letters 'E.H.G.' on one side of it and a single 'N' on the other" (*EH*, 11 August 1902). *Cabot* was subsequently laid up at Balaena for the winter (*EH*, 4 December 1902).

9 The northeast coast fishery began in late May, supervised by Adolph Nielsen (*EH*, 16 May 1902).

10 *EH*, 1 July 1902.

11 *EH*, 10 and 11 July 1902.

12 *EH*, 19 August 1902.

13 *EH*, 9 and 10 October 1902.

14 *EH*, 3 December 1902. There may have been some confusion in interpreting the act. It simply stated that "the holder of any such [whaling] licence shall not use, employ or operate more than one whaling steamer within the boundaries defined therein" Newfoundland, *Acts of the General Assembly of Newfoundland Passed in the 2nd Year of the*

Reign of His Majesty King Edward VII, 1902). It did not appear to prohibit a company owning multiple catchers, one for use at each station.

15 *EH*, 2 and 4 December 1902. *Viking* was sold to the Cape Broyle Whaling and Trading Co. after the one-year grace period given to the company expired.

16 *EH*, 28 July 1902.

17 *EH*, 6 September 1902.

18 *EH*, 18 and 21 August 1902. Between March and June, factory and guano plant construction materials and equipment were shipped to the site (*EH*, 28 May, 4 June 1902).

19 *EH*, 8 September 1902.

20 Ibid. Rissmüller had also built guano plants at Snook's Arm and Balaena, where in May 1902 "quite a number of men [were] employed and recently took 19 casks, each containing 60 gallons of oil from the whale handled" (*EH*, 31 May 1902). This carcass oil would have been of lower quality than that produced directly from blubber. Four grades of purity were identified at that time: 0–1 from blubber only, 2 from tongues, kidney fat, and residues from blubber boiling, 3 from flesh and bones and 4 from refuse (T.E. Salvesen 1914).

21 *EH*, 16 December 1902.

22 *EH*, 4 December 1902.

23 *EH*, 21 June 1902.

24 *EH*, 19 August 1902.

25 Tønnessen and Johnsen 1982, 104. The Ellefsen family were reputedly the owners of the harpoon patent in litigation between the Cabot and Newfoundland Steam Whaling companies. *EH*, 3 September, 14 November, 9 December 1902.

26 *EH*, 26 October 1901.

27 *EH*, 8 and 19 April 1902.

28 *EH*, 30 April 1902. Coal also arrived from Cape Breton on the *Helen Isabel*, and oil tanks from Horwood Lumber Co. in St John's (*EH*, 22 and 31 May 1902).

29 Tønnessen and Johnsen (1982, 104) identify Ellefsen's operation as the Atlantic Whale Manufacturing Co. That company was not incorporated until 22 February 1904 as an enterprise of the St John's merchant company Job Bros. It later bought the Aquaforte site (Atlantic Whaling and Manufacturing Co. Ltd 1904).

30 *EH*, 7 June 1902.

31 Tønnessen and Johnsen 1982, 104. Norwegian masters were also catcher gunners.

32 *EH*, 27 June 1902. Two humpbacks were killed on the first voyage.

33 *EH*, 3 and 20 September 1902.

34 *EH*, 27 December 1902.

35 *EH*, 4 July 1902. Whale lengths were often exaggerated. See, for example, *EH*, 11 July 1902. Blues killed from Balaena averaged some seventy-five feet (True 1904, 155).

36 *EH*, 2 August 1902.

37 *EH*, 27 December 1902.

38 Millais reported 472 whales (1907, 163). The species composition of the catch is unrecorded, although Mitchell and Reeves (1983, 166) estimate 105 humpback whales, extrapolated from the catch of thirty-seven out of the total kill of 108 from Aquaforte.

39 *EH*, 12 and 17 June, 19 July, 6 and 23 September 1902. In the second half of the nineteenth century, Scottish textile manufacturers concentrated in Dundee had converted to jute processing, a procedure requiring large quantities of whale and seal oil (Sanger 1980, 231–52).

40 *EH*, 8 October 1902.

41 The ongoing successes experienced by the three operating companies stimulated two

further licence applications in the last quarter of 1902. The directors of the Colonial Manufacturing Co. of St John's now proposed the establishment of factories at Cat Cove and Safe Harbour, both in Bonavista Bay (*EH,* 10 September, 28 October 1902).

42 *EH,* 5 and 7 January, 21 and 25 March 1903.

43 *EH,* 26 January 1903.

44 *EH,* 21 and 30 March 1903. The catch included at least eleven high-yielding blue whales. One reportedly ran for 122 miles and twenty-six hours before being killed with a second harpoon (*EH,* 31 March 1903).

45 *EH,* 21 May 1903. *Puma* was then transferred to the company's other station at Reuben's Cove.

46 The schooner *Robert T. Mason* brought oil barrels and general supplies to Reuben's Cove in late May (*EH,* 28 May 1903).

47 *EH,* 21 May, 2 June 1903.

48 *EH,* 8 October 1903.

49 *EH,* 27 July 1903. A later report indicated that ninety-two had been killed (*EH,* 24 August 1903).

50 *EH,* 14 August, 19 October 1903. By early September no less than 242 whales had been killed (*EH,* 3 September 1903).

51 *EH,* 13 November 1903.

52 *EH,* 9 November 1903.

53 *EH,* 4 February 1904.

54 See, for example, *EH,* 19 and 21 May, 19 and 30 June, 4, 6, and 7 July, 24 August, 19 September 1903.

55 *EH,* 17 September 1903. Rissmüller also chartered the barquentine *Antoinette* of St Pierre to collect oil and guano from Rose-au-Rue and Snook's Arm (*EH,* 23 and 24 October 1903).

56 *EH,* 23 April 1903.

57 *EH,* 31 March, 22 July 1903.

58 *EH,* 8 December 1903.

59 Ibid.

60 For details, see *EH,* 30 March, 15 April, 16 June, 6 and 27 July, 15 September, 16 November, 14 December 1903.

61 Baker et al. 1992, 31. See also *EH,* 2 November 1903.

62 Other members of the team included W. Palmer and J.W. Stollick (Sanger and Dickinson 1997, 79).

63 Thirty-three crates were shipped by sea to New York ("Summary of Costs for Getting the Sulfurbottom Skeleton for St Louis"). For a full account of the process, see *EH,* 11 July 1903.

64 Sanger and Dickinson 1997, 80–1.

65 The Board of Regents noted in its annual report for 1905 that "the cast, appropriately coloured, is now suspended from steel trusses connected with the framework of the roof in this Mammal Hall" (Smithsonian Institution 1906, 36).

66 Lucus 1933, 27.

67 For full details, see Sanger and Dickinson 1997, 81–4. See also Andrews 1929; *ET,* 13 May 1907; R.C. Murphy 1929.

68 *EH,* 7 January 1903. For details of machinery and supplies shipped to the site from St John's, see *EH,* 19 and 20 January, 10 and 28 February, 6 March 1903.

69 *EH,* 28 January 1903.

70 *EH,* 13 and 18 May 1903.

71 *EH*, 25 February 1903. Jacobsen's first job was to break up the bay ice so that *Viking* could supply the station (*EH*, 26 and 28 February 1903).
72 *EH*, 11 March 1903.
73 *EH*, 14 and 16 March 1903.
74 *EH*, 28 April 1903. *Viking* caught four blue whales, three humpbacks, and one fin.
75 *EH*, 19 June 1903. *Viking* killed up to five whales daily from a reportedly abundant stock (*EH*, 2, 6, 8 and 12 June 1903).
76 *EH*, 7 January 1903.
77 *EH*, 4 September 1903.
78 *EH*, 7 July, 9, 26 November 1903.
79 *EH*, 17, 18 and 23 July 1903. Although the thirty occupants lost most of their belongings, there were no fatal injuries, and some insurance was carried.
80 Ellefsen married Edith Melvin of St John's at the Anglican Cathedral on 5 January 1903 (*EH*, 5 January 1903).
81 *EH*, 9 January 1903. Work continued despite these incidents, with oil casks and lumber arriving on the *Trixie H.* and *Ethie* (*EH*, 6 and 7 February 1903).
82 *EH*, 7 February 1903.
83 Whales were "quiet and easily obtained" (*EH*, 16 June 1903) and up to thirteen caught per week, including four humpbacks and one blue in a day, "the best work for one day in the history of the fishery" (*EH*, 18 May 1903).
84 *EH*, 14 May 1903.
85 Newfoundland, *Annual [Fisheries] Report for 1903*, 1904, 4.
86 *EH*, 25 November 1903. A later report placed the catch at 217. *EH*, 18 December 1903.
87 *EH*, 8 October 1903.
88 The opening of the Cape Charles station reintroduced commercial whaling to coastal Labrador almost three centuries after the Basque had left.
89 *EH*, 18 September 1903.
90 *EH*, 1 October 1903.
91 *EH*, 28 March, 1 April 1903.
92 An original lease and licence held by Cashin were transferred by him to the company on 15 July 1904 in return for forty-five $100 shares.
93 *EH*, 22 December 1903.
94 *EH*, 2 September 1903.
95 *EH*, 18 July 1903.
96 *EH*, 19 October 1903.
97 Ibid.
98 *EH*, 9 November 1903.
99 *EH*, 6 October 1903.
100 *EH*, 19 October, 16 December 1903.
101 *EH*, 9 November, 16 December 1903.
102 *EH*, 14 December 1903.
103 *EH*, 16 December 1903.
104 Labrador Whaling and Manufacturing Co. Ltd, 1904. Ryan was also a shareholder in the St Mary's Whaling and Trading Co.
105 Trinity Whaling and Manufacturing Co. Ltd, 1904.
106 *EH*, 4 February 1903.
107 *EH*, 27 November 1903.
108 *EH*, 12 November 1903.
109 Harbour Grace Whaling Co. Ltd, 1904. Several shareholders (e.g., Charles Dawe,

W.D. Reid, and Dr L. Rissmüller) were involved in other companies.

110 Newfoundland, *Annual [Fisheries] Report for 1903*, 1904, 8.

111 *EH,* 30 March 1903.

112 *EH,* 15 April 1903.

113 See, for example, *EH,* 29 April, 4 June, 8 September, 12 October, 15 December 1903.

CHAPTER FIVE

1 Some of the participants also began to express concern that "the fishery would, inside of twenty years be a thing of the past" (*EH,* 30 March 1903). For a fuller account of the 1903 season, see Dickinson and Sanger 1993a.

2 Newfoundland, *Annual [Fisheries] Report for 1903*, 1904, 9.

3 *EH,* 26 January 1904.

4 *EH,* 4 February 1904. The company's catcher, *Puma*, killed 260 whales in 1903.

5 *EH,* 26 February 1904.

6 *EH,* 2 March 1904.

7 *EH,* 28 September 1904. This decision followed discussions with the Secretary of State for Scotland on the status of the whaling industry there and in Norway. There was widespread scepticism about the ability of the inquiry to provide an objective assessment since several government members were also major shareholders in whaling companies. *EH,* 4 October 1904.

8 *EH,* 12 August 1904.

9 Other reports indicated that this was not the case. The cod fishery adjacent to stations at L'Anse au Loup and Cape Charles in Labrador, for example, "ha[d] been the best known for years" (*EH,* 23 September 1904).

10 *EH,* 9 September 1904.

11 The declining Davis Strait and associated fisheries provided only ten whales in 1904, and although "whalebone and oil sell high, and a small quantity makes a profitable voyage … the boats will need to do better than they have done this year if Dundee is to keep whale fishing as one of its industries" (*EH,* 10 November 1904). For additional information on the decline of Dundee whaling, see Sanger 1980, 1988, and 1991d.

12 *EH,* 5 March 1904.

13 Ibid.

14 *EH,* 22 March, 12 April, 12 September 1904. This cargo was later described as over twenty thousand sacks of guano (*EH,* 16 September 1904). Forty-five tons were also sent to Halifax on *Fashoda* from the Cabot Steam Whaling station at Snook's Arm (*EH,* 16 March 1904). In addition, opportunistic collections were made. The schooner *Shamrock* arrived at St John's from Renews on 18 May 1904 "with 12 tons of whales bones. These were taken from the coves along the shore and were part of the carcasses of fish sent adrift from the factories before the fertilizing plant was established to dispose of the bodies of the fish killed by the whalers" (*EH,* 19 May 1904).

15 *EH,* 30 April 1904.

16 *New York Fishing Gazette*, reprinted in *EH,* 10 August 1904.

17 The seven factories operating in 1903 were capitalized at $350,000 and generated $10,500 in licence fees for the government (*EH,* 14 March 1904).

18 *EH,* 8 March 1904.

19 *EH,* 5 April 1904. Smith was a Scot who had been for many years the Dundee Sealing and Whaling Company manager in St John's. See, for example, *EH,* 28 July 1904. He

was subsequently "connected with the various whale factories here" (*EH*, 5 May 1903), including the new station at Dublin Cove (*EH*, 6 March 1903).

20 *EH*, 19 July 1904.

21 *EH*, 9 and 19 March 1904.

22 *EH*, 26 May 1904. Several Norwegians involved in the Newfoundland industry appear to have seen little future in their domestic fishery and proposed settling in Newfoundland (*EH*, 28 October 1904).

23 *EH*, 1 June 1904. The end of Finnmark whaling was partly a political act to prevent residents appealing to Russia for relief in the face of their declining cod fishery. Finnmark was adjacent to Finland, then Russian territory (*EH*, 1 July 1904).

24 *EH*, 22 April 1904.

25 *EH*, 25 April, 16 August 1904. Neither operated that year, and a station was never built at Conche.

26 *EH*, 12 May 1904.

27 *EH*, 11 May 1904.

28 *EH*, 31 May 1904.

29 *EH*, 23 September 1904. Newfoundland involvement in the development of shore-station whaling in British Columbia is discussed fully in Sanger and Dickinson 1991b and 1991c. See also chapter 8.

30 See, for example, *EH*, 6 and 9 January, 4 February 1904.

31 *EH*, 7 December 1904. In addition, "the price of whale oil is now quoted at £12 stg. a tun, and no sale at that. This is a surprising drop, for last year at this time it found a ready market at £24, but for some reason right this year oils have been unmarketable. It is thought by those interested that some mineral oils are taking the place of the whale product and as they can be obtained so much cheaper it has caused a slump" (*EH*, 13 December 1904).

32 This was valued at £2,250 to £2,800 per ton (*DA*, 3 November 1902, 21 October, 16 November 1903). In 1904, according to *Collier's Weekly*, "two and a quarter tons of whalebone ... sold in Dundee for $15,000 a ton, and only four tons are left outstanding in the world's supply" (reported in *EH*, 5 March 1904).

33 *EH*, 29 November 1904.

34 *EH*, 16 March, 27 May, 3 and 7 June, 2 July 1904.

35 *EH*, 1 August, 16 September, 16, 18, and 21 November, 29 December 1904.

36 *EH*, 7 and 14 May, 3 and 18 June, 6, 8, and 12 July, 9 September 1904.

37 *EH*, 27 June, 1, 2, and 15 August, 6 and 15 October, 12 and 16 November 1904.

38 *EH*, 28 January, 22 and 29 March, 2, 5, and 27 May, 14 June, 27 August, 13 September, 10 November 1904.

39 *EH*, 8 and 16 April 1904.

40 *EH*, 6 and 30 April, 27 May, 4 and 28 June, 16 and 22 September, 5 and 12 November 1904. *Falcon* took forty-five whales by the first week in June, eighty-five by the end of the month, and 133 to 22 September.

41 *EH*, 8 July, 1, 20, 22, and 30 August, 6 and 22 September 1904. Some of the bone was also shipped to Glasgow in September on the French barquentine *Paul Marie*.

42 *EH*, 19 and 29 April 1904. All did not proceed without problems, however, since "twenty men arrived from last night's train from Carbonear to go down to the new whale factory at Cape Charles. They were much disappointed because they were not met at the station by an agent of the concern, as several of them had no boarding houses and had not the wherewithal to procure food and shelter" (*EH*, 30 April 1904).

43 *EH*, 21 April, 16, 19, and 27 May, 14 June 1904.

44 *EH,* 9 and 19 July 1904.
45 *EH,* 23 and 26 July, 23 and 24 August, 23 September, 11, 15, and 24 October 1904. A catch of ninety-eight was recorded by the end of October.
46 *EH,* 12 November 1904.
47 *EH,* 22 and 23 November 1904.
48 *EH,* 16, 20, 26, 27, 28 and 31 May, 1 and 20 June 1904. A guano plant was also built at Riverhead for Dr Rissmüller.
49 *EH,* 30 June 1904.
50 *EH,* 24 August, 2 September, 4 and 17 November 1904.
51 For a fuller overview of Trinity's involvement in shore-station whaling, see Sanger and Dickinson 1998.
52 *EH,* 1, 14 and 25 June, 11, 15, 22 and 25 July, 6 August, 9 and 26 September, 7 and 28 November, 28 and 30 December 1904.
53 *EH,* 27 June, 16 July, 22 August, 6 September, 14, 27 October, 28, 30 December 1904.
54 *EH,* 5, 25, 26, 27 and 28 May, 1, 4, 11, 14 and 27 June 1904. Before *Mic Mac* ran aground at Port-aux-Basques, "two schooners off that port came across a large whale floating on the water dead with a harpoon imbedded in it attached to which was part of a steel cable. It was a splendid prize and the vessels' crews will dispose of it to some of the factories" (*EH,* 27 May 1904).
55 *EH,* 29 July, 1 August, 3 and 23 September, 15 October, 10, 15, 19 and 28 December 1904. *Mic Mac* was refloated "without any great damage" and returned to St John's via Dublin Cove.
56 *EH,* 3 and 6 May 1904. It had been argued before the Newfoundland Supreme Court the previous year that Rissmüller's station violated the requirement in the 1902 whaling act that stations must be at least fifty miles apart (*EH,* 28 October 1903). The Newfoundland Steam Whaling Company immediately launched an appeal to the Privy Council (*EH,* 29 October 1903). For full details, see PRO/CO 194/252 and 880/17.
57 *EH,* 31 October, 16 December 1903.
58 *EH,* 27 May, 4, 7, 16, and 28 June, 4 July, 16 September, 31 October 1904. Sixty-one packages of "whale material" arrived in St John's on *Mary Hough* on 6 June.
59 See, for example, the public debate and calls for international whaling regulations that appeared in the press following the close of the 1916 whaling season (*EH,* 27 November 1916).

CHAPTER SIX

1 Newfoundland, *Annual [Fisheries] Report for 1905*, 1906. All the catchers used in Newfoundland and Labrador up to 1905 were newly built in Norway by Nylands Mekaniske Verksted and Aker Mekaniske Verksted. See, for example, Job 1903; Job Brothers to Andreas Ellefsen, 27 May 1903; Northfield to Ellefsen, 6 December 1907. For additional details on Nylands's relationship with Newfoundland and Labrador whaling, see the twelve-page document, Aquaforte Whaling Co., *Contracts*, n.d. Catchers were also manned increasingly by more experienced crews. Local seamen were rehired from season to season, while the Norwegian masters, officers, and engineers usually returned on an annual basis, thus gaining additional knowledge of local conditions. For a fuller discussion of the decline that occurred in Newfoundland and Labrador shore-station whaling, see Sanger and Dickinson 2000.
2 The Neptune Steam Whaling Company was incorporated in 1904 with a public subscription of $90,000. A new catcher, *Neptune*, was built in Norway and work began on

the factory at Lark Harbour. Operations were suspended, however, "by reason of the unsettled Treaty Coast question." By the time the French shore problem was resolved, whaling had taken a downturn, and the investors decided not to proceed (*EH*, 28 September 1905).

3 *EH*, 24 March, 14 April 1905.
4 Ellefsen, Anders, to R. Berg, 21 April 1905.
5 *EH*, 17, 20 May 1905.
6 *EH*, 29 May 1905. The opening of the season further north was annually delayed by stormy seas and ice. By this date, 226 whales were killed: *Lynx*, forty-six, *Mic Mac*, thirty-eight, *Cabot*, thirty-eight, *Falcon*, twenty-four, *St Lawrence*, twenty-one, *Puma*, twenty, and *Avalon*, twelve.
7 *EH*, 24 June 1905.
8 *EH*, 5, 10 and 11 July 1905.
9 *EH*, 4 November 1905. Another report declared that "the collapse this fall in the whaling industry is provoking much discussion. There are 17 factories in operation altogether, and they represent an investment – for steamers, factories and guano plants – of fully a million and a half dollars almost every cent of which is local capital" (*EH*, 16 November 1905).
10 Newfoundland, *Annual [Fisheries] Report for 1905*, 1906.
11 Tønnessen and Johnsen 1982, 107.
12 *EH*, 16 November 1905.
13 *EH*, 20 December 1905.
14 Ibid. In a letter to the editor of the *Evening Herald*, signed "whaler," which was a response to an earlier article on a request by investors of "The Nail Factory" for tax relief, the correspondent asked: "Why should they receive any protection? There are other industries … that are giving ten or fifteen times more employment to the labouring class … in which local capital is also invested, with no protection. I refer to the whaling industry … I believe that some of the managing directors applied to the Government to have the licence fee either reduced or taken off, but they met with poor encouragement … why tax the whale fishery at all, any more than any other fishery?" (*EH*, 18 January 1906).
15 The premier, Sir Robert Bond, in responding to the industry's petition for relief, stated "that he would bring the matter before his executive colleagues … [but] could not, of course, intimate what their action would be as the whole subject would require careful consideration, but … was afraid there would be no prospect of remitting the present year's fees as they were provided by law and the Government [has] no power to cancel them" (*EH*, 20 December 1905). The government had decided initially to take no action (*EH*, 22 December 1905).
16 See, for example, *EH*, 20 October, 21 November, 4, 5, 19 and 27 December 1905.
17 "The Mic Mac Whaling Co., which operated the whaler of that name and factories at Dublin Cove and Lance au Loup [sic], but which had to voluntarily liquidate and reconstruct, being $27,000 in debt, met last week and decided to amalgamate with the Safe Harbour Whaling Company … The Phoenix will likely sell the '*Micmac*' [sic] and work Dublin Cove and Safe Harbour with the '*Fin*,' closing up Lance au Loup" (*EH*, 7 February 1906). For additional details, see, *EH*, 14 and 15 November, 15, 19, and 28 December 1906.
18 Newfoundland, Annual [Fisheries] Report for 1905, 1906, 150. For additional details of closures and amalgamations at the end of the 1905 season, see *EH*, 14 August, 3 October, 16 November, 20 December 1905.
19 *EH*, 14 August 1905.

20 The stations at Snook's Arm, Trinity, Lark Harbour, and L'Anse-au-Loup were closed.
21 "The whale fishery is a blank" (*EH*, 4 September 1906).
22 *EH*, 18 May 1906.
23 Colonial Cordage Co. to Anders Ellefsen, 19 March 1906.
24 See, for example, *EH*, 3 September 1906.
25 Ellefsen, Anders, to Colonial Cordage Co., 7 August 1906.
26 Munroe, J.H., to Anders Ellefsen, 10 August 1906.
27 *EH*, 17 April 1906. The delay appears to have been a conscious decision to reduce operating costs. Another indication of the dismal prospects for the 1906 season was provided in a letter received by Anders Ellefsen from a cooper he had employed previously. He ended by expressing "hope you have some whales by this time [March]. I would have sent before only I heard there was nothing doing" Carter, John, to Anders Ellefsen, 7 March 1906.
28 See, for example, *EH*, 12 and 30 April, 3 May 1906. In commenting on the scarcity of whales, the same newspaper reported: "Fishermen coming in from all parts of the northward were often met in schools right along the shore a few years ago. Many attribute … [the scarcity] to the constant hunting of the fish by the steam whalers now hunting here" (*EH*, 18 May 1906).
29 *EH*, 8 May, 29 June 1906.
30 *EH*, 26 May 1906.
31 *EH*, 18 June 1906.
32 Ibid. The *Hawk*, operating out of Cape Broyle, for example, "was as far as Grand Bank and this day reached Placentia Bay in quest of fish, but did not even sight a whale" (*EH*, 3 May 1906). Similarly, the *Lynx* (Reuben's Cove) "had to go 90 miles away from the coast" (*EH*, 1 June 1906). The *Baccalieu* (Harbour Grace) had returned "clean" following a three day cruise: "She was down to Bonavista Bay" (*EH*, 2 June 1906). The catchers at Cape Charles and Hawke's Bay, meanwhile, had "to go 100 miles to sea in their quest and report cetaceans are very scarce" (*EH*, 10 August 1906). See also *EH*, 18 June 1906.
33 *EH*, 29 June 1906.
34 *EH*, 25 July 1906.
35 *EH*, 14 July 1906. A further indication of the dismal state of the industry in general and Ellefsen's operation in particular is provided in a comment by John Maclellan, his oil and bone agent in Glasgow: "I am sorry the fishery has been so poor this season" (Maclellan to Anders Ellefsen, 6 July 1906).
36 *EH*, 20 September 1906.
37 See, for example, *EH*, 28 August, 2, 15 and 24 October, 7, 10, 15, and 17 November 1906.
38 Job Brothers to Anders Ellefsen, 23 August 1906.
39 Ellefsen, Anders, to Job Brothers, 25 August 1906; Job Brothers to Anders Ellefsen, 27 September 1906.
40 *EH*, 19 April 1906.
41 *EH*, 15 September 1906.
42 *EH*, 4 December 1906.
43 The "southern" and "western" fisheries were complete failures. In late April a St John's newspaper reported that "people who recently arrived from the south coast say that there is an abnormal scarcity of whales on the Newfoundland shore, and that whalers say that not a fish has been seen for weeks past between Cape Spear and Cape Ray. The fish now at Chaleur factory were taken off the Cape Breton coast" (*EH*, 22 April 1907).

Conditions had improved little by the end of May: "For the past week nothing has been done by the western whale factories" (*EH,* 30 May 1907).

44 For example, *Hump* went from Dublin Cove to the station at Trinity in late July "to cruise northern Bays … [where] she captured twelve of the big fish between Catalina and Cape Bonavista" (*EH,* 9 August 1907). *Cabot* was just as successful at Snook's Arm, landing "over 30 fish" by mid-September (*EH,* 19 September 1907). It was also reported that "this has been the most successful year since the station at Beaverton [*Lynx*] opened" (*EH,* 4 November 1907). *Cachalot* did "even better at Hawke Harbour, managing to catch 63 whales, which is considered a good season's work" (*EH,* 23 November 1907).

45 *EH,* 13 January, 10 February 1908.

46 *EH,* 28 August 1908.

47 *EH,* 26 October 1908.

48 Once again whales in 1908 were "not plentiful along the south coast" (*EH,* 8 May 1908). See also ibid. 3, 15 and 18 June 1908. Similarly, the lone west coast factory at Hawke's Bay had not processed a single whale by the end of June, "and the prospects are very poor, the fish having all passed down [north] through the straits [of Belle Isle]" (*EH,* 24 June, 28 August, 26 October 1908). The Hawke's Bay station closed in September "after a very unsuccessful year" (*EH,* 25 September 1908). For a full account of the latter half of the 1908 whaling season, see *EH,* 6, 23, and 24 July, 11, 26, 27, and 28 August, 1, 5 12, 19, 22, and 23 September, 2, 6, 8, 14, 23, and 26 October, 3, 9, 10, 12, 13, 21, and 24 November, 9, 12, 15, 17, and 23 December 1908.

49 Their delivery voyages were not always without incident. *Fin*, for example, left St John's for Kobe on 4 January 1907 with *Avalon* and *Neptune*, reached the Azores three weeks later, then went into the Mediterranean and through the Suez Canal. On 25 February it ran aground near Jeddah in the Red Sea and four of the crew were lost. For full details, see *EH,* 4, 5 and 10 January, 19 and 21 February, 13 March 1907.

50 See, for example, *EH,* 4 January, 13 May 1907.

51 An editorial on the industry's "outlook" noted that "there will be a big shortage … not more than 7 steamers operating against 16 last year, but the men who are not employed in this pursuit are being readily absorbed by the cod fishery or by the Harmsworth enterprise at Grand Falls" (*EH,* 20 April 1907).

52 The first shore-station started operations in November 1904. It was established at Grytviken, South Georgia, by the Compañia Argentina de Pesca (Hart 2001). The first floating factory vessel was located at the South Shetland Islands in 1905 (*Report of the Interdepartmental Committee on Research and Development in the Dependencies of the Falkland Islands* 1920).

53 Also, the brother of Captain Nokard Davidsen, the Newfoundland Steam Whaling Company's whaling manager, was master of the catcher *Hauken*, attached to the floating factory *Admiralan* at Deception Island in the South Shetland Islands.

54 The company planned to deploy a factory ship to process southern whales in the austral spring and summer, then work the Newfoundland and Labrador coastal waters in the northern spring (*EH,* 29 April 1907).

55 *Report of the Interdepartmental Committee on Research and Development in the Dependencies of the Falkland Islands* 1920.

56 *Sobroan* was built at Sunderland, England, in 1889 (*Mercantile List and Maritime Directory*, 1907).

57 *EH,* 22 October 1907.

58 Tønnessen and Johnsen 1982, 106. Included was the construction of a side-slipway

designed after a Norwegian patent granted in 1905 to Captain Nokard Davidsen, whaling master for the Newfoundland Steam Whaling Company (Johnsen 1947).

59 *EH,* 23 October, 6 November 1907.

60 Crew List, *Sobraon* 1907.

61 *EH,* 19 and 20 November 1907.

62 *EH,* 11 and 21 May 1908. Whales were seen at the South Orkney Islands but could not be caught because of heavy ice.

63 For full details, see Headland 1987; "Hvalfangernes Gravsteder i Antarktis": 9–38; Lang 1908. The graveyard was covered in debris following a volcanic eruption on the island in 1969.

64 Harrington 1987b.

65 *EH,* 14 March, 11 May, 25 July 1908.

66 Tønnesson and Johnsen 1982, 106–7.

67 Plans were made to refit *Sobroan*'s storage tanks in an attempt to prevent spoilage in future voyages (*EH,* 1 September 1908).

68 *EH,* 9 August 1910. See also Dickinson and Sanger 1991 and 1993b.

69 Rose-au-Rue Whaling Co. Ltd, 1910–15.

70 "This year the whaling catch will be one of the biggest on record, the average being 90 for each steamer" (*EH,* 4 November 1909).

71 The one exception on the island was Placentia Bay which continued to be a favourite haunt of the depleted stock of migrating whales. Only Hawke Harbour, however, the most northerly station, produced consistently profitable returns. See, for example, *EH,* 13 June 1912.

72 After Anders Ellefsen's attempts to continue whaling failed in 1907, he again placed advertisements in local newspapers (*EH,* 3 July 1908) and contacted prospective buyers in the United States, South Africa, Nova Scotia, and other parts of Canada. In his opinion, Anders indicated, whales locally "should come back" (Aquaforte Whaling Co., Letterbook, 14 May 1908). Job Bros. was given first chance to purchase the station because of its "guano factory there [at Aquaforte]" (Ellefsen, Anders, to Job Brothers, 2 April 1908). Anders also considered that an additional incentive for the purchase of his factory was its "salvage" potential because of Aquaforte's proximity to the notorious Cape Race, where "when [a] steamer is out whaling, she has [a] good chance of being first at the scene of the wreck" (Ellefsen, Anders, to Joseph Salter's Sons, North Sydney, N.S., 9 July 1908). No serious offers were received, however, and Anders wrote to his father in Norway, "Cannot obtain on offer. May I leave?" (Ellefsen, Anders, to Andreas Ellefsen, 10 October 1908). For a full account of the rise and decline of the Ellefsen whaling operation at Aquaforte, see chapter 7.

73 See, for example, *EH,* 20, 28, and 30 March, 1, 6, 8, 11, 13, and 18 April 1910; 6, 8, 10, 13, 14, and 16 January 1914; 4, 11, and 19 February, 1, 3, 6, 9, and 12 April 1915; 2 February, 4 May, 19 September 1916.

74 The collapse of the Newfoundland and Labrador whaling industry was temporarily delayed by rising oil and bone prices after World War I (*EH,* 18 April 1914; 2 May 1915).

75 For seasonal details of the Cabot Steam Whale Co. Ltd during 1909–14, see *EH,* 6 and 24 March, 8 and 12 June, 3 and 17 July, 24 August, 7 October, 30 November, 13 December 1909; 25 January, 22 April, 13 May, 4 July, 9 and 16 September, 25 and 26 October, 2, 5, and 22 November 1910; 16, 20, and 24 March, 29 April, 8, 20 May, 15 and 21 June, 1, 11, 18, and 22 July, 4 August, 26 September, 4 and 5 October, 6 and 9 November, 6 December 1911; 8 and 10 April, 14, 16, and 17 May, 13 and 15 June, 19 July, 2, 12 and 16 August, 5 and 25 September, 1, 5, 8, and 14 October, 13 November 1912; 4 February, 19 March, 1, 9, 12, and 28 April, 8 and 31 May, 27 June, 7, 8, 17, and 18 July,

14, 16, and 29 August, 12 September, 21 November 1913; 4 February, 23 and 27 March, 25 April, 13 and 21 May, 6, 16, and 23 June, 28 July, 3 and 9 August, 29 September 1914.

76 For seasonal details of the Newfoundland Steam Whaling Co. Ltd during 1909–15, see *EH,* 22 February, 19 and 23 March, 3, 12, 13, 16, 20, and 21 July, 24 August, 3, 17, and 18 September, 6 and 22 October, 2, 3 and 5 November 1909; 28 March, 9, 20 and 30 April, 5 and 27 May, 7 June, 1, 18 and 26 July, 24 August, 13 and 16 September, 22, 28, 29 and 30 November, 6 December 1910; 20 March, 17 May, 21 June, 31 July, 7, 19 August, 8 September, 5, 11, 17, 19, 21, and 28 October, 25 and 27 November, 6 December 1911; 1 and 3 April, 3, 17, 21, and 22 May, 13, 21 and 25 June, 15, 25 and 26 August, 9 October, 13 and 18 November 1912; 12 April, 8 and 31 May, 26 June 1913; 4 February, 16, 23 and 29 June, 22 August, 8 and 29 September 1914; 19 March, 16 and 24 April, 12 May, 17 June, 25 October, 1915.

77 For seasonal details of the Cape Broyle Whaling and Trading Co. Ltd during 1909–12, see *EH,* 23 March, 12 June, 21 August, 3, 7, and 18 September, 18 and 29 October, 16 and 22 November 1909; 1, 10, and 19 March, 13 and 20 April, 23 May, 8, 22 July, 6 August, 3 and 16 September, 1, 15, and 31 October, 5, 28, and 29 November 1910; 20 March, 7 April, 10, 13, 16, 17, 18, and 25 May, 1, 8,10, 15, 16, 20, 21, and 27 June, 4 and 24 July, 22 and 24 August, 12 and 30 September, 5 and 14 October, 6 December 1911; 17 April, 8, 17, 21, and 25 May, 8, 10 and 20 June, 2, 12, and 17 August, 2 September, 13 November 1912.

78 For seasonal details of the Phoenix Whaling and Manufacturing Co. Ltd during 1909–14, see *EH,* 13, 24 March, 1, 3, and 5 April, 3, 12, and 30 July, 24 August, 2, 3, 14, 16, 17, 18, and 28 September, 11 October, 16, 18, and 30 November, 1 December 1909; 25 January, 23 March, 8, 9, 12 and 14 April, 3, 4, 5 and 10 May, 17, 24 and 29 June, 1, 8, 11, 14, 22, and 30 July, 10, 15, 20, and 22 August, 1, 2, 6, 8, 13, 16, 20, 22, 27, and 29 September, 1, 4, 7, 14, 17, 18, and 25 October, 15, 22, 28, 29, and 30 November, 1, 2 and 3 December 1910; 16, 20, and 24 March, 8, 11 and 22 April, 8, 16 and 18 May, 6, 12, 15, 21, and 27 June, 7, 8, 10, 18 and 25 July, 25 and 26 September, 14 October, 2, 7, and 25 November, 6 December 1911; 20, 21, 28, and 30 March, 13 April, 13, 17 and 21 May, 5 and 18 June, 19 July, 20 August, 30 September, 11 October, 12 and 13 November 1912; 27 January, 3, 12 and 21 April, 8 and 14 May, 10, 12 and 13 June, 7, 8, 12, and 29 July, 9 August, 1, 24 and 30 October 1913; 4 February, 7 and 23 March, 25 and 28 April, 21 and 30 May, 6, 16, and 23 June, 29 September, 7 October 1914.

79 For seasonal details of the Hawke's Bay Whaling and Manufacturing Co. Ltd during 1909–13, see *EH,* 19 May, 9 July, 21 August, 3 and 18 September, 1 and 9 October 1909; 18 April, 9, 13, 15, 16, 17, 22, 23, 24, 28 and 30 September, 1, 4, and 7 October, 2, 28 and 29 November 1910; 20 March, 26 April, 18, 22 and 24 May, 8 and 9 June, 18 and 29 August, 12, 28, and 29 September, 5 October, 30 November, 2 and 6 December 1911; 9, 16 and 17 May, 6 and 13 November 1912; 28 April, 8 May 1913.

80 For seasonal details of the Labrador Whaling and Manufacturing Co. Ltd during 1909–10, see *EH,* 17 and 24 May, 21 June, 12, 20, 22 and 28 July, 10, 30 August, 3 and 15 September, 6, 12 and 26 October, 3, 10, 16, 17, and 18 November 1909; 17 and 23 May, 23 June, 4, 23 and 25 July, 10, 22, 26, and 30 August, 7 and 19 September, 18, 22, and 30 November 1910.

81 "Word was received Saturday [27 August] that the whaling steamer *Cachalot* was beached at Crowhead Cove [eastern end of Belle Isle] and is not likely to hold together much longer as the bottom is almost completely pounded out of her. The fishing season is now at its best and the loss to the Company is a serious one … as whale fishing on that coast [southern Labrador] does not finish until late in the fall" (*EH,* 29 August 1910). See also *EH,* 25 August, 2 and 7 September 1910.

82 "She is the same size as the old *Cachalot* but is stronger in the bow and amid ships. She is also fitted with the latest appliances and the most modern gun" (*EH,* 6 June 1911).

83 For seasonal details for the Labrador Whaling and Manufacturing Co. Ltd during 1911–16, see *EH,* 4, 11 and 25 May, 8 and 28 June, 16 and 31 August, 1 September, 3 and 5 October, 2 and 30 November, 1, 6, and 7 December 1911; 23 January, 9, 29 and 30 April, 14 and 20 May, 5 June, 16 and 19 July, 3 and 30 August, 5 October, 13 and 29 November, 9 and 13 December 1912; 24 January, 28 March, 8, 12, 17, 29 and 31 May, 30 June, 8 and 30 July, 16 and 21 August, 5, 12, 15 and 24 September, 18, 20 and 24 November 1913; 23 January, 13, 21, 30 May, 6, 16, 23 June, 2, 17 July, 19, 27 August, 16, 28, and 29 September, 22 and 24 October, 3, 16, 20 and 23 November 1914; 15, 16 and 24 April, 2, 8, 12, 17, and 26 June, 3, 8, 17, 20 and 23 July, 7, 10, and 28 August, 13, 18 and 25 September, 13, 14, 15, 20, and 21 October, 2, 6, 10, and 25 November, 1, 2, 3 and 10 December 1915; 14, 15, 17, and 20 January, 22 and 30 May, 2, 7 and 27 June, 1 and 18 July, 1, 5, 9, 19, 21, and 29 August, 4 and 6 September, 2, 24 and 27 October, 13, 16, and 24 November 1916.

84 *EH,* 12 May 1907. *Cachalot II* was subsequently chartered for "patrol work."

85 *EH,* 28 July 1917.

CHAPTER SEVEN

1 For a more extensive examination of Ellefsen participation in Newfoundland and Labrador whaling, see Dickinson and Sanger 2001. See also Dickinson and Sanger 1993b.

2 Tønnessen and Johnsen 1982, 31.

3 For a full discussion of Foyn's role in the development of Finnmark whaling, see, for example, Niemi 1993.

4 Tønnessen and Johnsen 1982, 25. The Ellefsen family, and their involvement in whaling in general, has been described by Falnes 1999, 81–100.

5 Wogn-Henriksen 1983–84. Their vessels included the barques *Alma, Fridtjof, Oscar,* and *Petrus,* and the brig *Ellef Torsen,* all built at Skjaersnes.

6 Tønnessen and Johnsen 1982, 71. "The Norwegian government has prohibited whale fishing for a period of ten years, beginning February 1, 1905, and the sum of $100,000 has been appointed to indemnify the proprietors of whaling establishments for the loss sustained by this prohibition" (*EH,* 13 January 1905).

7 See, for example, *EH,* 25 and 27 June, 5, 23, and 29, July 1901.

8 *EH,* 26 October 1901.

9 The company was incorporated in Norway with shares issued at Kr 15,000 each (circa $3,750), and the capital raised was transferred to twenty-six-year-old Anders Ellefsen, Andreas's second eldest son. This was for his use as construction superintendent and operations manager of the station, to be repaid at 5 per cent interest (Aquaforte Whaling Co., Shareholder Agreement, 30 October 1901). Anders had trained as a gold/silversmith and worked in Tønsberg, Oslo, Germany, Spain, and the United States before becoming involved in the family's Finnmark whaling operation (Wogn-Henriksen 1983–84, 19).

10 Clift, Thomas B., [lawyer, St John's] to Andreas Ellefsen, 5 December 1906; *EC,* 21, 25 and 28 May 1908; *EH,* 5 December 1906; Atlantic Whaling and Manufacturing Co. Ltd, 1912.

11 *Humber* was ordered at Kr 79,000 for delivery in April 1902. She was completed on time, and the company transferred her to Anders Ellefsen on 13 May 1902 for

Kr 85,000, "delivery before arrival in Newfoundland" (Aquaforte Whaling Co., Contract Agreement, 16 November 1901, 13 May 1902, and 7 June 1902). *Humber* was Nyland hull no. 132, and was 90′x 17′10″ x 11′2,″ with a capacity of 280 gross tons. The price included "all equipment."

12 *EH,* 31 October, 6 November 1901. See also *EH,* 30 October 1901.

13 *EH,* 31 October, 1901.

14 Colonial Secretary to C. O'Neill Conroy, 11 October 1901. At that time the Newfoundland government was debating legislation to regulate the new industry in what was to become the 1902 whaling act.

15 *Act to Regulate the Whaling Industry*, 1902, 103.

16 *EH,* 8, 12 February, 21, 22 and 24 March, 16 and 19 April, 22 and 31 May 1902. Wooden oil tanks were built by the Horwood Lumber Company, St John's. Coal and other supplies arrived on the barque *Helen Isabel* from New Campbellton, N.B. There seems to have been confusion as to whether or not Anders Ellefsen had a local agent for the *Humber*. The *Evening Herald* (11 June 1902) recorded that A.G. Prowse held this position, whereas on 14 June 1902 it reported that "Mr. Ellefsen has appointed no agent to date and will transact the business of the company here personally."

17 *EH,* 19 and 23 April, 9, 11, 14, 18, and 27 June, 4 and 11 July, 3 and 20 September, 27 December 1902.

18 *EH,* 3 September 1902.

19 In an attempt to reduce these discrepancies, legislation was passed that prohibited Norwegians from crewing catchers after a station had been operating for three years. The act, however, did not significantly alter the division of labour and status, although one local captain was the master of *Hawk* in 1911. The previous season, for example, the master of the *Hump* was paid $2,260 for seven months' work, and the crew received $300 each. By contrast, station workers earned only about $25 a month (*ET,* 1 December 1910; 20 March 1911). See also Aquaforte Whaling Co., Employment and Wage Ledger, 1902–03; Dickinson and Sanger 2001, 46–7.

20 *EH,* 30 August, 5 and 6 September, 20 October, 22 November, 5 December 1902. For jute processing in Scotland, a major user of Newfoundland seal and whale oil, see Sanger 1980, 231–52.

21 *EH,* 22 September 1902. See also *EH,* 20 August, 20 September 1902. A blue whale captured in late August was "valued at $700." *Humber* also hunted off the north coast of the Avalon Peninsula due to the different seasonal migration patterns of individual species. A total catch of ninety-five was recorded.

22 *EH,* 8, 9, and 27 December 1902. Another account reported 103 killed at the beginning of the month. Sanger and Dickinson 1991c, 16.

23 B. Melvin, St John's, pers. com., 1993 (related to the Kiserud family via Eda Hopkins Melvin).

24 Marriage certificate, Anders Ellefsen and Eda Hopkins Melvin, St John's, 1903 (estate of Marie Kiserud); Marie Kiserud, Norway, pers. com., 1991 (granddaughter of Eda Hopkins Melvin and Anders Ellefsen).

25 James Melvin subsequently worked at Aquaforte in 1903 for $10 per month. Also a witness was Eda's best friend, Linda Martin, who visited her in Norway just prior to World War I. Eda did not revisit Newfoundland following her move to Norway, which, according to her daughter, Marie, was very difficult for her (M. Kiserud, pers. com., 1991).

26 *EH,* 9 January 1903. His St John's lawyer wrote: "I was very sorry to hear of your fire, which occurred at a most unfortunate time. However, you were lucky to save the factory, which would have been a very serious loss" (Conroy, C. O'Neill, to Anders Ellefsen, 18 February 1903).

27 *EH,* 10 and 15 January, 3 and 6 February 1903. The visit was not without incident. A Norwegian crewman had his hand crushed while *Humber* was tying up in St John's. He was treated by Dr J.W. McGulloch, a witness at the wedding. The new house was completed on 5 April (*EH,* 6 April 1903).

28 In early June "the great fish were there [off Aquaforte] about in schools" (*EH,* 3 June 1903). For full details of the early portion of the 1903 whaling season, see *EH,* 6, 13, 14 and 18 May, 3 June, 2 July 1903.

29 *EH,* 17 and 18 September, 8 and 12 October, 25 November, 18 and 22 December 1903; 9 and 28 January, 4 and 15 February 1904.

30 Aquaforte Whaling Co., Whale and Oil Ledger, 1902–08. The *Evening Herald,* (29 November 1903) recorded a catch of 230 whales.

31 *Act to Regulate the Whaling Industry*, 1902, 102.

32 Ellefsen, Anders, to Dr L. Rissmüller, 4 May 1903.

33 Newfoundland, *Annual [Fisheries] Report for 1902*, 1903, 4. The impact of flensed carcasses disposed of at sea from Aquaforte is evidenced by an American banking schooner that "ran into a carcase [sic] last month [August], at night, and the crew thought they had struck a submerged wreck until the odour assailed them. 'I can smell it yet', remarked the captain, describing the incident a fortnight later" (*EH,* 16 September 1903).

34 The government's whaling inspector, C. Way, for example, claimed privately that "as far as I have been able to ascertain these carcasses must certainly come from your factory at Aquaforte" (Way, C., to Anders Ellefsen, 3 September 1903).

35 Way, C., to Anders Ellefsen, 19 September 1903.

36 Conroy, C. O'Neill, to Anders Ellefsen, 5 September 1903.

37 Ellefsen, Anders, to Dr L. Rissmüller, 4 May 1903. Rissmüller agreed to pay $24 for blue whale carcasses, $14 for fins, and $10 for humpbacks.

38 Conroy, C. O'Neill, to Anders Ellefsen, 10 October 1903.

39 Hunter, J.R., to Anders Ellefsen, 26 May 1903.

40 Rissmüller, Dr L., to Anders Ellefsen, 5 June 1903.

41 Hunter, J.R., to Anders Ellefsen, 3 June 1903. Hunter later agreed to purchase all the bone, "if you can get Rissmuller [sic] to let you keep the jaws" (Hunter, J.R., to Anders Ellefsen, 30 June 1903).

42 Rissmüller, Dr L., to Anders Ellefsen, 16 June 1903.

43 See, for example, Hunter, J.R., to Anders Ellefsen, 16 July 1903; Ellefsen, Anders, to J.R. Hunter, 24 July 1903; Huffman, W.F., to Anders Ellefsen, 29 July 1903; Huffman Manufacturing Co. to Anders Ellefsen, 13 August 1903.

44 Huffman, W.F., to Anders Ellefsen, 23 October 1903.

45 Conroy, C. O'Neill, to Anders Ellefsen," 10 October 1903.

46 Job Bros., Memorandum of Agreement, 13 February 1904.

47 *EH,* 22 March 1904.

48 Henrik also married a Newfoundlander, Hett Cross, of Aquaforte. They returned to St John's in the *Bruce* on 19 March (*EH,* 20 March 1905). The station had not been completely abandoned during their absence, as the wife of Norwegian blacksmith Hans Hang died there on 17 February. The body was "embalmed by undertaker Carnell [St John's] and sent to Norway for interment" (*EH,* 18 February 1905).

49 Ellefsen, Anders, to Andreas Ellefsen, 21 April 1905. His position is perhaps best indicated by the fact that in the same letter he explains that he has just purchased a typewriter "from my own money of course, couldn't be that extravagant with the registered company's money."

50 Ellefsen, Anders, to R. Berg, 21 April 1905. He was forced to buy barrels on "3-month

accepts, I have only paid cash for about 300 out of the total of approximately 1500 barrels. About $800.00 in the bank in addition I am expecting roughly $900.00 from Philadelphia and about $300.00 in petty cash. I would have been able to pay cash for more barrels as you can see but was reluctant to make myself short of cash and then have to telegraph too early after money. If only you were able to sell the oil as soon as it arrived so that we could get in some money."

51 *EH,* 19 April 1905. For details of the beginning of the 1905 season, see *EH,* 24 and 27 March, 5, 12, and 14, April 1905. *Humber* also transported oil and bone procured in 1904 to St John's, in one instance "having in tow a small schooner laden with whale products." Other supplies, such as oil casks, arrived at the station on Job's new catcher *Fin.*

52 *EH,* 12, 13, 15, 20, and 24 June, 5, 10, and 11 July 1905.

53 *Humber,* for example, "was out from Aquaforte for two days and got back ... without a fish. She did not see one on the cruise" (*EH,* 2 August 1905).

54 Newfoundland, *Annual [Fisheries] Report for 1905,* 1906, 150. Detailed monthly statistics kept by the company show that ninety-two whales (thirty blue, twenty-eight fin, and thirty-four humpbacks) were killed between April and October 1905 (Aquaforte Whaling Co., Letter Book, 18 January 1906). See also *EH,* 6 and 14 October 1905.

55 *EH,* 8 May, 18, 19 and 29 June, 14 and 25 July, 4 September 1906. An indication of the dismal state of the industry, and Ellefsen's operation in particular, is provided in a comment from Ellefsen's Glasgow oil and bone agent: "I am sorry that the fishery has been so poor this season" (Maclellan, J., to Anders Ellefsen, 6 July 1906).

56 Berg, R. to Anders Ellefsen, 17 July 1906.

57 Ibid. The Aquaforte station, in fact, was closed when the 1906 catch produced "a loss so resounding that it wiped out the profits of the four previous years" (Tønnessen and Johnsen 1982, 108). Quite clearly the Ellefsens had learned from their earlier whaling experiences at Finnmark and Iceland.

58 Payne, W.B., to Anders Ellefsen, 11 September 1906.

59 Ellefsen, Andreas, to Anders Ellefsen, 2 November 1906.

60 *EH,* 5 November 1906.

61 *EH,* 21 November 1906. See also Rankin, C.B., to Anders Ellefsen, 5 December 1906.

62 Ellefsen, Anders, to T. Clift, 24 November 1906.

63 By Rankin's account the factory and plant cost $3,150, 721 casks $1,000, and 15 tons of coal $60. Offset against this amount was $12.00 for advertisements in three newspapers and Cliff's 2 per cent commission of $84.20 (Rankin, C.B., to Anders Ellefsen, 5 December 1906).

64 *EH,* 27 September 1906. After numerous crew changes and delays, *Humber* finally left Newfoundland on 4 October 1906 (*EH,* 5 October 1906).

65 Robert Shankland and Co. to Anders Ellefsen, 20 December 1906.

66 Bryde, Johan, to Anders Ellefsen, 8 November 1906.

67 Ellefsen, Arne, to Anders Ellefsen, 22 March 1907; Ellefsen, Andreas, to Anders Ellefsen, 2 October 1906.

68 The willingness of the two brothers to go to Japan illustrates the entrepreneurial spirit that initially brought the Ellefsens to Newfoundland. Ragnvald Berg, in a letter exploring the possibility of Henrik taking *Humber* to Japan, for example, asked Anders to "say hello to Henrik and Hett [his Newfoundland wife] – I am almost afraid of what the latter has to say about Henrik going to Japan" (Berg, R., to Anders Ellefsen, 17 July 1906).

69 Length 77 feet; breadth 15 feet; depth 11 feet; sloop-rigged; gross tonnage 73.88 (Ellefsen, Anders, to Harvey and Company [St John's], 1 August 1908).

70 Ellefsen, Andreas to Anders Ellefsen, 12 December 1906.

71 *EH,* 26 December 1906.

72 *EH*, 19 and 24 August, 6 September 1907. Aquaforte Whaling Co., Letter Book, 12 September 1907.
73 In an effort to make ends meet, for example, the Ellefsens had extended their "agricultural" operation, attempting to sell eggs to Water Street (St John's) stores (Rankin, C.B., to Anders Ellefsen, 11 December 1907; Rankin, C.B., to Anders Ellefsen, 21 September 1908).
74 *EH*, 3 July 1908.
75 See, for example, Ellefsen, Anders, to Joseph Salters [North Sydney, N.S.], 11 July 1908; Ellefsen, Anders, to P.F. Windsorton [Durham, South Africa], 25 July 1908; Aquaforte Whaling Co., Letter Book, 14 May 1908.
76 Ellefsen, Anders, to Job Brothers, 2 April 1908.
77 Brister and Son [Halifax] to Anders Ellefsen, 25 June 1908.
78 See, for example, *EC*, 21, 25, and 28 May 1908.
79 Ellefsen, Anders, to Andreas Ellefsen, 10 October 1908.
80 Ellefsen, Andreas, to Anders Ellefsen, 20 October 1908.
81 B. Melvin, St John's, pers. com., 1993.
82 *EH*, 13 November 1908.
83 Ellefsen, Anders, to Andreas Ellefsen, 10 October 1908.
84 "Christensen, C.L., in Memoriam," 1951, 7–9.
85 Kosmos, A/S, and Anders Ellefsen, Agreement, 18 August 1911.
86 Cook, Tasker, [the lawyer who had arranged the sale of the Aquaforte whaling station to A/S Kosmos], to H.M. Maddick, 4 January 1930. He indicated that "this company is out of business since 1912."
87 B. and E. Kiserud, Norway, pers. com., 1991 (sons of Marie Ellefsen Kiserud).

CHAPTER EIGHT

1 Whaling in British Columbia and the "Pacific Northwest" is well documented in Hagelund 1987; Tønnessen and Johnsen 1982, 111–46; R.L Webb, 1984, 1988a, and 1988b.
2 R.L. Webb 1984, 3.
3 The first whaler to enter the South Pacific was the British vessel *Amelia*, "manned by the Nantucket colony of whalemen; and sailing for that ocean from London in 1787, her first mate, Archelus Hammond, killing the first sperm whale known to have been taken in that ocean" (Starbuck 1878, 90). In 1835 *Ganges* (Captain Folger) of Nantucket, opened up the Kodiak ground (Clark 1887, 17).
4 For fuller details of Newfoundland involvement in B.C. whaling, see Sanger and Dickinson 1991a and 1991b.
5 G. Balcom, Vancouver, pers. com., 2003 (descendant of Sprott Balcom). Sprott Balcom's whaling-sealing and West-East Coast connections have also been described by his son Lawrence H. Balcom (Hagelund 1981, 56–65) and R.L. Webb 1988b, 146–56.
6 Eastern Canadian involvement in South Atlantic sealing is discussed in Dickinson 1989.
7 As R.L. Webb (1988b, 148) reports, "he engaged the interests of his old colleagues at the Victoria Sealing Company, particularly William Grant, who, like the Balcoms, had owned and commanded sealing schooners."
8 *Orion* was built for the company in Christiania, Norway, in 1904. She was renamed *Pluvius* in 1930, converted, and served as Vancouver's fireboat until 1945 (Hagelund 1982, 198). Her arrival at Victoria in February 1905 generated considerable interest in Newfoundland. See, for example, *EH*, 24 February 1905.

9 R.L. Webb 1984, 9. Similar dividends were paid in Newfoundland and Labrador.

10 Page's Lagoon was intended to be an off-season station and provide winter employment for boat crews and shore workers from the other two factories. Unfortunately, whales were scarce, and the plant closed after one season. Thereafter whaling in British Columbia would remain a seasonal occupation, requiring that the previous season's employees be called up each spring (R.L. Webb 1988b, 166–70). See also Hagelund 1982; Tønnessen and Johnsen 1982.

11 R.L. Webb 1988b, 178.

12 Ibid., 192.

13 See Hagelund 1982, 66–79; and R.L. Webb 1984 and 1988b for information provided by Schupp's grandson, William Lagen.

14 R.L. Webb 1984, 13.

15 Ibid. See also R.L. Webb 1988b, 238. The industry ceased temporarily at the end of the 1943 season when the Consolidated Whaling Corporation stopped operating.

16 Ibid. 1984, 16, and 1988b, 256–77.

17 Taiyo Gyogyo Kabukishi Kaisha's involvement in whaling is outlined in Tønnessen and Johnsen 1982, 135–46. Japan's involvement in British Columbia whaling is described in R.L. Webb 1988b, 278–87.

18 R.L. Webb 1984, 16.

19 Correspondence of 11 July 1904 between R.E. Finn, Sprott Balcom's lawyer in Halifax, and J.-R.F. Préfontaine, Canadian minister of Marine and Fisheries, for example, states that Captain Balcom would be "leaving in a day or two" to examine the Newfoundland fishery. On 15 July 1904 Balcom was listed on the *Bruce* from North Sydney to Port-aux-Basques but not on the passenger list for the train arriving in St John's (*DN*, 16 July 1904). He probably only travelled as far as Little St Lawrence. Not only was Dr Rissmüller, the owner/operator of the Little St Lawrence whaling station, favourably disposed to receiving visitors (Millais 1907) but when he visited Balcom at Victoria in March of the following year, it was at the invitation of "an old acquaintance" (*EH*, 30 March 1905).

20 For fuller details, see, for example, Sanger and Dickinson 1991b, 100–3.

21 Family connections between the Balcoms in Halifax and Victoria are provided in Hagelund 1982, 56; Tønnessen and Johnsen 1982, 114–15; R.L. Webb 1988b, 146–56.

22 Reported in *EH*, 5 August 1898.

23 R.L. Webb 1988b, 147.

24 Finn, R.E., to J.-R.F. Préfontaine, 11 July 1904.

25 *EH*, 9 November 1903.

26 Finn, R.E., to J.-R.F. Préfontaine, 29 July 1904.

27 *EH*, 25 July 1904. "The others" included Edward (Ned) Scaplen, a marine and mechanical engineer who had also been heavily involved in the construction and operation of whaling stations in Newfoundland (O. Miles [Scaplen's daughter], Victoria, B.C. pers. com., 1990).

28 *DN*, 28 July 1904.

29 *EH*, 22 July 1904. Charles Smith had for many years been manager of the Dundee Sealing and Whaling Company operations in St John's. (*DN*, 28 July 1904).

30 An example of the entrepreneurial drive of those involved in Newfoundland whaling is provided in the Canadian minister of Marine and Fisheries' response to an investor in Sydney, N.S. who wanted "rights" to whale along the Cape Breton coast. Préfontaine noted that there had "been several applications for such, principally from Newfoundland" (Préfontaine, J.-R.F., to A. Johnson, 14 November 1903).

31 They were unsuccessful primarily because of B.C. legislation that restricted whaling station construction "by any person except residents five years in the province" (*EH*, 23

September 1904; Préfontaine, J.-R.F., to A. Walker," 14 November 1904).

32 In response to a request for information from J. Pope, under-secretary of state, for example, the deputy minister, Marine and Fisheries, pointed out that the minister, J.-R. F. Préfontaine, "intends to introduce a Bill next season authorizing the licensing of whale factories, and following in many details, the system legalized in Newfoundland" (Gourdeau, F, to J. Pope, 26 November 1903). See also Préfontaine, J.-R.F., to A. Johnson, 14 November 1903.

33 There were minor differences. The annual licence fee in Newfoundland at that date, for example, was $1,500, compared to $1,200 in B.C. Similarly, in response to a specific request, the Canadian act specified that only harpoons and lines could be used in the hunt. For further details, see Sanger and Dickinson 1991c, 103–4.

34 R.L. Webb 1988b, 151.

35 *VC* 16 November 1916. Reported in *EH,* 24 November 1916.

36 Tønnessen and Johnsen 1982, 105; R.L. Webb 1988b, 156. Both appear to rely upon Millais 1907.

37 Part of his estate was left "to Dr. Julius, who lived in Germany" (*EH,* 24 December 1916).

38 His wife, Pauline, claimed after his death that her father "was one of the richest men in New York" (ibid).

39 "He was a traveller who provided transient addresses on important legal documents" (R.L. Webb 1988b, 158).

40 *EH,* 20 September 1900.

41 McCourt, n.d., 8; also reported in R.L. Webb 1988b, 157.

42 P.W. Brown 1909, 168.

43 *EH,* 19 August, 7 December 1901.

44 *ET,* 28 May 1902. Not all held him in such high esteem. Olive Miles, a girl of seven when Rissmüller visited Page's Lagoon in the winter of 1907–08, for example, recalled him as "being proper" (O. Miles, Victoria, B.C., pers. com., 1990). Sprott Balcom's son, Lawrence, thought "he was the biggest blowhard of all, had more wild ideas. Always trying to promote something" (Hagelund 1982, 60). Interestingly, his background was in plant science (Rissmüller 1873).

45 "Dr. Rismuller [sic] has started in to manufacture oil at Balaena from the offal carcasses of the whales taken there treating them chemically and being highly successful in procuring good liquid" (*EH,* 31 May 1902). Reports from Rose-au-Rue also indicated that the guano factory there was "in full operation" with the whales "being put through processes which convert them into an excellent fertilizer" (*EH,* 13 August 1902). One month later the plant was "operating full blast, the machinery is working splendidly" (*EH,* 8 September 1902).

46 R.L. Webb 1988b, 157, refers to Cape Broyle as "Cape Royal."

47 *EH,* 29 September 1902.

48 Conroy, C. O'Neill, [solicitor for the Colonial Manufacturing Company] to T.J. Murphy [Newfoundland minister of Marine and Fisheries], 18 November 1902.

49 *EH,* 9 November 1903. See also *EH,* 18 July 1903.

50 Millais 1907, 184.

51 The substance of these patents is described in R.L. Webb 1988b, 158–61.

52 Notary Public Document, 6 January 1905. He also unsuccessfully negotiated with French authorities to relocate his station from Little St Lawrence to St Pierre off Newfoundland's Burin Peninsula (fig. 1.1) in return for a twenty-year licence monopoly (*EH,* 16 August 1905).

53 R.L. Webb 1988b, 156.

54 *EH,* 13 December 1905. This was actually his second visit that year as he was reported to be at Sechart just before it opened, at the invitation of Sprott Balcom, "an old acquaintance" (*EH,* 30 March 1905).

55 R.L. Webb 1988b, 157.

56 Balcom, S., to R.P. Venning [asst. commissioner of Fisheries], 2 September 1907. This point has been accepted by subsequent observers. In Newfoundland, however, the facts do not support Tønnessen and Johnsen's contention that "when he came to Newfoundland (in 1900) the whaling industry was not in a favourable state and was showing no profit, but once Rismuller [sic] entered the business, things soon changed" (1982, 106). R.L. Webb (1988b, 161) claims that in British Columbia the "'Rissmuller Patented Whale Reduction Process' appears to have been the only means of meeting Ottawa's stringent processing prerequisite."

57 R.L. Webb 1988b, 180. In 1909 Sprott Balcom was listed as managing director, with Dr L. Rissmüller, director (Pacific Whaling Co. Ltd, 1909). Dr Rissmüller is listed as vice-president and general manager, and Captain Sprott Balcom as second vice-president (*EH,* 20 December 1910).

58 According to Hagelund 1982, 48, Rissmüller "pursued his dream of an American whaling company by securing, through a holding company in Seattle, a site at Westport in Gray's Harbour." See also R.L. Webb 1988b, 189.

59 *VC,* 16 November 1916; *EH,* 24 November 1916.

60 Goddard 1986, 209.

61 R.L. Webb 1988b, 156–69.

62 Huswick 1912, 1.

63 Goddard 1986, 209.

64 *EH,* 6 March 1903. Gosney managed the station at Riverhead until it closed. For details of Gosney's move to British Columbia, refer to Goddard's taped interview with his daughter, Nellie Marmo (Marmo n.d).

65 *EH,* 17 May 1909. In 1911 Peter Cashin, working at various jobs in western Canada as a youth, visited his "old friend 'Dickie' Carroll … [who] was working in a very responsible position with Dr. Rissmuller in the whaling company" (Cashin 1976, 139).

66 *EH,* 26 January 1911; R.L. Webb 1988b, 237. George Le Marquand's involvement with Newfoundland and British Columbia whaling is described in Goddard's taped interview with his son, Alan Le Marquand (Le Marquand n.d.). The four top managerial positions listed on Victoria Whaling Company letterhead in 1916 were occupied by men previously involved in the Newfoundland industry. S.C. Ruck was general manager, while the station managers for Bay City, Kyuquot, and Naden Harbour were listed as G. Le Marquand, C.E. Ruck, and A. Gosney, respectively (Victoria Whaling Co. Ltd, 1916).

67 Goddard 1986, 209.

68 Ibid.

69 O. Miles, Victoria, B.C., pers. com., 1990.

70 Hagelund 1982, 127.

71 *EH,* 4 August 1906.

72 *EH,* 31 July 1906.

73 Cashin 1976, 33. Cashin's detailed account of the 1912 whaling season provides an excellent indication of the roles Newfoundlanders continued to play in British Columbia whaling following the 1910 reorganization of the industry.

74 Ibid., 139.

75 Ibid., 141. For a full description of life on the British Columbia stations before World War I, see Goddard 1986; R.L. Webb 1988b.
76 Cashin 1976, 141.
77 *EH,* 8 July 1907.
78 *EH,* 30 July 1907.
79 *EH,* 11 August 1908. Three years later J. Thompson and J. Walsh, newcomers to the British Columbia industry, wrote to friends in Newfoundland that they were "perfectly satisfied with conditions of affairs as they exist in Victoria, B.C." (*EH,* 3 May 1911). Similarly, four ex-policemen from St John's also wrote home in 1911 with "optimistic reports." Although on "a lonely island" [in the Queen Charlottes], they were paid extra for overtime, and since there was "nothing else to do," they could earn "tall" money, as much as "$75 a month," whereas if they had stayed in the police force they would have been earning "only $15." There were "concerts and games," they continued, to occupy "their spare moments" (*EH,* 26 August 1911).
80 Goddard 1986, 209.
81 D.B. MacPherson 1985.
82 Ruck, S.D., [general manager, Victoria Whaling Co.] to Department of Immigration [Victoria], 6 April 1916.
83 Jas. Murphy and Sons: Letterhead, 1916.
84 Jas. Murphy and Sons to Victoria Whaling Co. Ltd, 17 May 1916. See also ibid., 11 September 1916; Victoria Whaling Company Ltd. to Jas. Murphy and Sons, 21 September 1916.
85 D.B. MacPherson 1985.
86 *EH,* 8 October 1906.
87 *EH,* 5 October 1906.
88 For details of the voyage, see *EH,* 23 October, 10 and 28 November, 24 December 1906; 4 and 12 February 1907.
89 *EH,* 25 November 1907.
90 *EH,* 26 November 1907.
91 The long association between *St Lawrence* and British Columbia whaling is described in Hagelund 1982, 198; Tønnessen and Johnsen 1982; and R.L. Webb 1988b.
92 Hagelund 1982, 198, records it only as the *Sebastian* in Newfoundland.
93 *EH,* 28 October 1909.
94 For details of *Sebastian's* voyage round Cape Horn to British Columbia, see *EH,* 30 November, 2 December 1909; 15 and 22 January, 1 and 17 February, 15 and 22 March 1910.
95 Hagelund 1982, 198. *Sebastian* was sold to the Canadian Northern Railway and converted into a towboat (R.L. Webb 1988b, 226).
96 Ibid., 261.

CHAPTER NINE

1 See, for example, *EH,* 5 June, 9 August 1918. The new operator was Chr. Hannevig (Mitchell 1974, 108–9).
2 *EH,* 9 August, 19 October 1918; 12 April 1919. For a fuller account of the renewal that occurred between 1918 and 1936, see Sanger and Dickinson 1995; between 1937 and 1951, see Dickinson and Sanger 1999b.
3 *EH,* 10 May, 16 June 1918. Sergeant 1953, 439-47 and Mitchell 1974 recorded the catch as 101. *The Daily News* (7 October 1918) reported fifty-one caught by *Port Saunders* and

forty-eight by *Hawk*. See also Dickinson and Sanger 1993b, 91–100.

4 For fuller details, see *PF* (various); *Norsk-Hvalfangst Tidende* (various); Tønnesen and Johnsen 1982. D. Vinnedge, Freemont, Calif., (pers. com., 2001) supplied further information.

5 *DN*, 8, 16 May, August 1919; *EH*, 11 September 1919. Operations were delayed by sea ice (*EH*, 25 August 1919), while "the prevalence of storms on the Labrador coast" hindered the end of the season (*EH*, 11 November 1919). Mitchell (1974, 133) records 1,464 barrels of oil produced. Two additional Labrador fishing properties were also placed on the market "owing to Fishery Regulations and Combines" (*EH*, 2 March 1920).

6 *EH*, 15 June 1920. It should be noted that all the evidence refutes Ryan's intention. At this point, in fact, he was vigorously promoting the revival of the local industry so that he could recover his investments in the inactive Hawke Harbour Station.

7 The "dwelling house" at Cape Broyle was subsequently destroyed by fire in February 1921 (*ET*, 3 February 1921).

8 *ET*, 21 March, 5, 7, 10, and 17 August, 12 September, 30 December 1922.

9 *ET*, 18 January 1923; *DN*, 18 January 1923.

10 *ET*, 13 February 1923. Beaverton was also referred to as Beaver Cove. The stations at Lark Harbour, Aquaforte, and Cape Broyle did not reopen.

11 Sorvald, Håkon, to C.W. Sanger, 22 December 1992; British Norwegian Whaling Co. Ltd, 1926–34. Equipment was brought to Grady on the Norwegian vessel *Herlof*, "taken possession of an hour after her arrival [in St John's, en route to Norway] at the instance of the ship SS *Morelos*, her owners, masters and crew … on a claim for assistance alleged to have been rendered in Labrador" (*ET*, 19 November 1927). See also *Hvalfangernes Assuranceførening Gjensidig, 1911–36*; *ET*, 11 September 1928; 16 August 1934. In 1927, 3,313 barrels of oil valued at $82,823 were produced (*ET*, 4 January 1928).

12 The Cape Broyle Whaling and Trading Co. Ltd and *Cabot* were sold to the new company for $150,000 in cash and shares (*ET*, 12, 17 May, 6 June 1923; 12 April 1924). *ET*, 18 January 1923, reported "there are evidences that whales are again putting in an appearance along our coast." For a list of shareholders, see the Newfoundland Whaling Co. Ltd, 1923–36.

13 See *ET*, 6 June, 5 July 1923. The Newfoundland Whaling Company purchased *Cachalot* from the Labrador Whaling and Manufacturing Company and bought *Cabot* at auction "fitted and ready for use" for $6,150 from Crosbie and Company (*DN*, 10 January 1922).

14 *ET*, 25 November 1924, reported seventy-one whales killed. See also *ET*, 23 October 1923. The oil was usually sent to Holland and Denmark on the company's SS *Gullaug* (Capt. Neilson), which normally returned with coal for the stations (*DN*, 24 November 1925; *ET*, 5 June, 1 November 1926; 9 May, 6 September 1927).

15 *ET*, 6 June 1922, carried an auction advertisement listing the articles and structures of Rose-au-Rue.

16 *DN*, 16 April 1924. *ET*, 14 November 1924, reported 183 whales killed (109 from Rose-au-Rue and seventy-four from Hawke Harbour). Another report gave a figure of 174, including sixty-five for *Cabot* (forty-two from Hawke Harbour and twenty-three from Rose-au-Rue) and seven at Hawke Harbour and 109 from Rose-au-Rue for *Cachalot* (*ET*, 25 November 1924). See also *ET*, 19 August 1924; 24 March, 20 April, 15 May, 22 June, 2 and 3 July, and 24 August 1925; *DN*, 22 June and 24 November 1925.

17 *DN*, 27 July 1927; 30 November 1928.

18 Whale oil exports from Newfoundland and Labrador over the eight years to June 1928 were valued at $892,100; 1920–21, $45,852; 1921–22, $303; 1922–23, $4,206; 1923–24, $36,471; 1924–25, $87,324; 1925–26, $209,650; 1926–27, $214,543; 1927–28, $333,751 (*ET*,

19 May 1926; 20 August 1927; 12 June, 31 December 1928).

19 *ET,* 3 September 1927; 25 February, 10 April 1929; 21 May 1930. This decline was similar to that occurring in British Columbia, where "the catch of all species has waned considerably, although increasing efficiency of the hunters has led to an increase in catch in recent years." Global oil markets had become so saturated by the early 1930s that Norwegian companies agreed not to catch in 1931–32, "a black day in the history of Norwegian whaling" (Tønnessen and Johnsen 1982, 391). For full details, see ibid., 386–94; G. Jackson 1978, 204–12.

20 *ET,* 20 April 1931.

21 *ET,* 23 July, 16 November 1931.

22 A correspondent to a local paper suggested that if the company were not going to reopen, government should operate the stations to provide employment. When hunting resumed, a "recently purchased" tanker, SS *Moss* (Capt. Thorald Colwyn), was used (*ET,* 6 May, 4, 17, and 18 July, 4, 9, 11, and 21 August, 12 and 20 October, 23 November, 11 and 13 December 1933).

23 *ET,* 16 May 1933.

24 *ET,* 23, 30 November, 5 December 1933. Wages of $50,000 were paid.

25 Newfoundland Whaling Co. Ltd, 1923–36; Anonsen, Amund, to Chr. Salvesen Co. Ltd, 31 December 1936; Newfoundland Whaling Co. Ltd, Directors' Minutes, 16 January 1934. Anonsen worked as operations manager at $3,000 per annum, plus production bonuses of thirty cents per barrel of oil, ten cents per two-hundred weight sack of guano and 5 per cent of net profits.

26 Oil and guano were exported in 1934 from Rose-au-Rue to Liverpool on SS *Havtor* of the Newfoundland-Canada Steamship Line and on SS *Vardfjell,* and to Canada on SS *Hernanda.* The next year one thousand tons of oil were shipped from Rose-au-Rue and Hawke Harbour to Denmark on SS *Moira* (*ET,* 21 April, 11, 14, and 23 May, 9, 21, and 26 June, 19 July, 15 and 16 August, 6 and 27 September, 19 October, 18 and 28 November 1934; 8 and 30 January, 6 February, 30 March, 4, 13, 25, and 27 June, 5 and 13 July, 8 October, 9 and 30 November 1935). Olsen was knighted in 1935 by the King of Norway. The company was recorded as still owning the inactive properties at Beaverton and Cape Broyle. In 1936 the board of directors asked the Newfoundland Fisheries Board "to set up a plant to manufacture fish oil and fish meal, a protein-rich product that [Crosbie] believed would be suitable for livestock and poultry feed" (M. Harris, 1992, 147). Both Crosbie and Olsen later became partners in the Olsen Sealing and Whaling Co. Ltd.

27 H.K. Salvesen, 1935. The draft agreement was signed by Olaf Olsen and received in April 1935. The Salvesen family came to Scotland from Norway, and through Chr. Salvesen and Co. subsequently operated various whaling and other companies, including the South Georgia Co., the Olna Whaling Co., the Sevilla Whaling Co., and the Polar Whaling Co. (List of the Archives of Messrs. Christian Salvesen Limited). See also Elliot 1998; Vamplew 1975.

28 Salvesen to A. Anonsen, 29 October 1936; Anonsen, A., to Salvesen, 31 December 1936; Anonsen, A., to Salvesen, 1 February 1937; Salvesen to A. Anonsen, 16 April 1937; Newfoundland Whaling Co. Ltd, Directors' Minutes, 9 March 1938. A total of 3,723 shares were held, "not including I.R.S. Salvesen's personal holdings" by Salvesen firms and family: South Georgia Co., 1,228; Chr. Salvesen and Co., 1,000; Capt. H.K. Salvesen, 500; N.G. Salvesen, 500; T.N.F. Salvesen, 495.

29 Their names are listed in the *Evening Telegram,* 15 October 1936.

30 *ET,* 14 and 27 March, 11 April, 26 May, 1, 5, 15, 24, and 25 June, 2, 3, 15, and 20 July, 28 November 1936; 31 March 1937.

31 According to H.K. Salvesen (n.d.), facilities at Hawke Harbour included twenty-four pressure boilers for cooking blubber, two separators for removing the oil, a guano plant, storage tanks for nine thousand barrels of oil, and space for one thousand tons of coal. The wharf had to be taken up every winter to prevent ice damage, and its condition and the shallow adjacent water meant that large transports were usually loaded and unloaded by barge while lying in the bay. The company purchased the hulk *Lobelia* from Bowring Bros. and sank it in the bay as a breakwater to improve entry (*ET,* 21 October 1930). See also *ET,* 8 April 1952.

32 *Signa* and *Spina* were built in Middlesborough in 1926 at £15,250 each and were both 189 gross tons, 105 feet in length, 22 feet in breadth, and 12 feet in depth. *Silva* was built for £16,000 in 1924, being 221 gross tons, 110 feet in length, 23 feet in breadth and 13 feet in depth (Salvesen to Ministry of Shipping, London, 6 August 1940). In Newfoundland and Labrador they were run through the Polar Whaling Company, a subsidiary that leased the catchers to the Newfoundland Whaling Company at approximately $1,000 each, plus transit expenses.

33 H.K. Salvesen, n.d. *Sevilla* was built in 1900 by J. Priestman and Co. of Sunderland for the Hamburg-America Line. She was interned in Argentina during World War I, converted to British registry in 1919, and used by Salvesen as a floating factory through their subsidiary A/S Sevilla until 1929. She was then sold to Anders Jahre of Sandefjord and repurchased by Salvesen in 1933 as a transport ship. *Sevilla* was finally sold in 1949 for breaking up in Belgium (Bakka, Jr 1992, 150).

34 Christian Salvesen Co., Letterbook. Salvesen was also the oil sales agent for Hans Ellefsen during his pre-World War I whaling operations in Iceland (Vamplew 1975, 166–7).

35 Christian Salvesen Co., Position of Ships Book. An engineer from the Newfoundland Whaling Company, James Cook, and six firemen and sailors were on board (*EH,* 15 March 1937). See also *ET,* 19 April, 4 June 1937.

36 Salvesen to Newfoundland Whaling Co. Ltd, 1 February 1937; Salvesen to Board of Trade, London, 26 May 1937; Christian Salvesen Co., Whaling Officers.

37 *ET,* 15, 26 October 1937. The season started on 5 June and ended on 15 November (*ET,* 15 July, 12 and 17 August, 15 and 26 October, 29 November 1937). The directors received $3,000 each, and the accountant $250. Anonsen's salary of $7,735 consisted of a basic $3,000 plus $3,840 bonus on 19,200 barrels of oil at twenty cents per barrel and $895 on 17,910 bags of guano at five cents each (one bag of guano = 2 cwt; 17,910 bags = 1,791 tons). The cost of producing one ton of oil was £13.6.1d (£17.1.7d per ton excluding depreciation) and £9.4.1d for one ton of guano (Newfoundland Whaling Co. Ltd, n.d,; Christian Salvesen Co., Hawke Harbour Operation, n.d). *Meta* took 292,530 gallons of oil to Rotterdam (*ET,* 29 November 1937). The catch included a "sea monster killed in Fortune Bay" by *Morelos,* later identified as a right whale, the first (and only) reported killed since local modern hunting began in 1898 (*ET,* 12 and 17 August 1937).

38 Anonsen, A., to Salvensen, 23 July 1937; Newfoundland Whaling Co. Ltd, Directors' Minutes, 1 April 1938.

39 Newfoundland Whaling Co. Ltd, Directors' Minutes, 6 April 1938; ibid., 10 May 1938; *ET,* 26 May, 10 June 1938.

40 In the Matter of the Companies Act and in the Matter of the Newfoundland Whaling Company Limited in Liquidations, n.d. *ET,* 12 August 1938, gives the bid as £19,000 for the stations and equipment and £39,000 for oil stored in Rotterdam.

41 Fox, Knight and Phelan [St John's] to Salvesen, 25 January 1940. The purchase price is variously recorded as £10,342, about $50,000 at prevailing exchange rates (Salvesen to Liquidator, Newfoundland Whaling Co. Ltd, 27 February 1940), or $300,000 including whale oil (*ET,* 12 August 1938). Salvesen appears to have been given permission in

November 1938 to buy the equipment, facilities, and oil (*ET*, 9 November 1938).

42 *Anonsen vs. The Newfoundland Whaling Company, 1941.*

43 A.H. Murray and Co. Ltd is identified as the agent for the Polar Whaling Co. Ltd in a letter from the Atlas Insurance Co. of Montreal enquiring about providing insurance for company employees at Harbor Grace (Atlas Insurance Co. to A.H. Murray and Co. Ltd, 28 December 1939). Catchers were laid up there for the winter. (Salvesen to A.H. Murray and Co. Ltd, 15 November 1939).

44 Newfoundland Whaling Co. Ltd, 1923–36.

45 Salvesen to Food (Defense Plans) Dept., London, 28 September 1938.

46 Christian Salvesen Co. Ltd, Notes, 8 October 1939. The sperm oil was sold to the Gorton Pew Fisheries of Gloucester, Mass., and to the Mardon-Wild Corp. of Somerville, Mass., at six cents per pound (A.H. Murray and Co. Ltd to Mardon-Wild, 27 May 1941; A.H. Murray and Co. Ltd to Gorton Pew, 30 May 1941).

47 Tønnessen and Johnson (1982, 457–71) discuss the problems facing the global whaling industry in 1937–39.

48 Christian Salvesen Co. Ltd, Hawke Harbour Operations, n.d.; Salvesen to Board of Trade, London, 13 October 1939; *ET*, 7 July, 2 November 1939; 16 January 1940. The station provided employment for two hundred. Four men from the Placentia area worked in South Georgia.

49 British whaling companies were brought under control of the Ministry of Food in 1939 as a war measure (Salvesen to Board of Trade, London, 13 October 1939).

50 A.H. Murray and Co. Ltd to Salvesen, 24 September 1940.

51 Salvesen to Ministry of Shipping, London, 8 February 1940. About 160 workers were employed at Hawke Harbour. Transport vessels such as *Sevilla* were also not requisitioned, since they were being used to carry oil important to the war effort. Salvesen proposed laying up their tankers *Sourabaya* and *Peder Bogen* in Newfoundland with full crews "until we could get employment for them rather than wait in dangerous waters at home," but this was not done and both were sunk by enemy action. Their iron ore carrier *Saganaga* also suffered the same fate, being torpedoed by a German submarine when loading at Bell Island, Newfoundland (Vamplew 1975, 117, 123; S. Neary 1994).

52 Salvesen to Ministry of Shipping, London, 10 February 1940.

53 Ministry of Shipping [London] to Salvesen, 7 February 1940.

54 Salvesen to Ministry of Shipping, 2 February 1940; ibid., 29 February 1940; ibid., 11 April 1940; Admiralty to Salvesen, 29 August 1940; Ministry of Shipping to Salvesen, 15 March 1940; A.H. Murray and Co. Ltd to Salvesen, 24 September 1940; Trading with the Enemy Department [London] to Salvesen, 20 July 1942; *ET*, 3 May 1941. A further twenty-seven catchers owned by Salvesen and either operating or laid up in the South Atlantic were also requisitioned for war duties. Borgan purchased *Rio Sama* largely from monies owed him by Salvesen.

55 A.H. Murray and Co. Ltd to Salvesen, 24 September 1940; *ET*, 12 October 1942. About 3,200 barrels of oil were on hand during September, when whaling was to be "discontinued soon."

56 Salvesen to Post and Telegraph Censorship Department, London, 13 May 1943.

57 *Gun VI* was originally *Subra*, built for Salvesen at Smith's Dock, Middlesborough, in 1916 (Polar Whaling Co. Ltd to General Registrar and Record Office of Shipping and Seamen, London, 24 June 1943). The vessel was sold in 1929, then bought by the Salvesen company A/S Sevilla in 1936. It was renamed *Gun 6* until purchased by the Polar Whaling Company for work in Labrador as *Gun VI*. It was resold again in 1946 (Vamplew 1975, 277).

58 Salvesen to Admiralty, London, 8 January 1943. *Charcot* was bought for £3,250 and

Gun VI for £1,500 (Polar Whaling Co. Ltd, New Account, 1930–47). Anders Herlofsen and Jacob Ness were master and chief engineer of *Charcot* (Christian Salvesen Co., Notes, 27 May 1943; ibid., 2 August 1943).

59 Admiralty to Salvesen, 11 June 1943.

60 Salvesen to Admiralty, London, 27 May 1943. Leith Harbour operated for only a short period in 1941–42. Otherwise, whaling from Grytviken was the only such activity in the Antarctic and sub-Antarctic from 1941 until January 1944 when a Norwegian expedition was despatched (Tønnessen and Johnson 1982, 483; Salvesen to Admiralty, London, 19 June 1943; Admiralty to Salvesen, 13 April 1944).

61 Salvesen to Admiralty, London, 15 February 1944. Layovers of eleven–twelve days were necessary in Rio de Janeiro and Pernambuco, respectively, as was a detour from Trinidad via Florida.

62 *ET*, 20 September 1945; 16 and 27 November, 9 December 1946.

63 Polar Whaling Co. Ltd, Ledger, 1930–47; Vamplew 1975, 278. Bakka 1982 records the fate of several ex-Salvesen vessels. *Sousa* may have been bought by Hvalfanger A/S Suderøy of Haugesund to work with their factory ship *Suderøy*. *Silva* was sold for £15,000.

64 *ET*, 13 and 25 July, 9 August 1947; 8 June 1948.

65 Oil was also transported by the tanker/supply vessel *Polar Maid* (Stevenson 1950–1). *Southern Gambler* was built in 1950 at Smith's Dock, Middlesborough, for Salvesen and hunted from Labrador during the summer of 1950 before going to Leith Harbour, South Georgia. It also worked for the Saldanha Whaling Co. during 1960–61, then as a buoy boat for the Salvesen factory vessel *Southern Harvester* in 1962–63. Laid up thereafter until 1966, *Southern Gambler* was converted into a fish factory ship prior to being sold to the Cayman Islands in 1988 (Bakka 1982, 184). *Southern Lily* was used as a towboat (*ET*, 11 May 1948). It also took twenty-one whales on the way from St John's to Hawke Harbor (*ET*, 20 October 1948). The presence of *Southern Laurel* is given in *ET*, 28 May 1949, *Soika* in *ET*, 8 June 1948, and *Southern Hunter* in *ET*, 9 June 1951. For the names of masters, see *ET*, 25 May 1948. A contemporary catcher, *Southern Actor*, is displayed at the Sandefjord Whaling Museum.

66 *ET*, 1 June 1949.

67 *ET*, 22 April, 1 May 1948.

68 Bakka 1982. According to R.L. Webb (1998b, 265), the Western Whaling Company of British Columbia bought *Bouvet III* in 1950. The vessel "had seen considerable Antarctic service since its construction in Middlesborough, England in 1930, and had most recently worked in Labrador" (*ET*, 23 November 1953).

69 For local comment on these vessels, see Flynn 2002. Additional information supplied by D. Flynn, Conception Harbour (pers. com. 2003) and A. Kuehn, Karlsruhe, Germany (pers. com. 2003).

70 *ET*, 25 May 1948. On 14 October 1950, the *Evening Herald* reported that stocks previously found near Hawke Harbor were "frequenting waters further offshore."

71 Stevenson 1950–1.

72 *ET*, 13 April 1950.

73 *ET*, 1 December 1949.

74 *ET*, 12 October 1948. These concerns were largely rejected (*ET*, 24 November 1948).

75 *ET*, 18 November 1950.

76 *ET*, 18 and 22 November 1950.

77 IWS, 1930–52.

78 *ET*, 8 April 1952. At least 136,936 barrels of oil and 5,935 tons of meal (guano) were produced (Vamplew 1975, 286).

79 Vamplew 1975, 245. Johan Carlson Borgen had also commanded catchers for Aristotle

Onassis's *Olympic Challenger* whaling fleet, which hunted in the Antarctic without regard for regulations established by the IWC. Borgen was a "huge man about 6′4″ tall; in 1952 when still working for Onassis, he was said to have killed more than 5,100 whales in twenty one years of shooting." His son, Arne Borgen, once gunner/master of *Shila,* was also one of the most influential *Olympic Challenger* gunners, and was asked to revamp the Western Whaling Co. operations from Coal Harbour, B.C. during the mid-1950s (R.L. Webb 1988b, 270, 362).

80 Salvesen continues to be active in transport, cold storage, and quick freezing from Scotland.

81 A.H. Murray and Co. Ltd to Salvesen, 23 October 1940; Olsen Whaling and Sealing Co. Ltd, 1940–45; Vamplew 1975, 230–1. *Morelos* was sold for $13,000 (Salvesen to Ministry of Shipping, London, 4 November 1940).

82 Olsen, O., to Salvesen, 28 May 1940. Olsen also expanded his interests by purchasing a St John's machinist company, Gulbrandson and Co., with "a view to handling all kinds of repairs on ships which put in here in distress" (*ET,* 26 November 1942).

83 *ET,* 17 May 1941; G. Lilly, St John's (company accountant and son of H.R. Lilly, former whaling manager at Hawke Harbour and Rose-au-Rue) pers. com., 1989. "Over $50,000 were spent in operations [in the first season, 1941] and all the men are satisfied with their pay" *(ET,* 27 October 1941).

84 In preference to alternate sites examined in Green Bay and Notre Dame Bay (*ET,* 21 September 1944).

85 *ET,* 26 February 1945.

86 *ET,* 21 September 1944.

87 *ET,* 13 October, 24 November, 24 and 27 December 1945; 5 February 1946.

88 *ET,* 30 May, 4 July 1945; 18, 22 January, 5 February 1946. The barge's crew had to be evacuated by breeches buoy, and *Morelos* was also driven out to sea in the gale.

89 *ET,* 1 February 1946.

90 *ET,* 31 January, 19, 20 and 21 November 1947. The vessel also rescued twenty-five wrecked fishermen from the schooners *M and G Fowlow* and *Scarlet Knight* in 1949 (*ET,* 15 October 1949). In recognition of this last rescue, the British government presented "the captain, mate and two A.B's … [with] a silver bowl, a pair of binoculars and two silver cigarette cases" (*ET,* 19 November 1949).

91 *ET,* 17 February, 15 May, 19 June 1947.

92 *ET,* 7 February 1947.

93 *ET,* 27 March 1947.

94 *ET,* 13 July 1948. Whale meat sold there for some $800 per ton (*ET,* 19 April 1949).

95 *ET,* 13 July 1948. Including the activities of Salvesen, this season was considered "one of the most lucrative on record" for the Newfoundland whaling industry as a whole *(ET,* 4 October 1948).

96 *ET,* 18, and 26 May 1949.

97 *Morelos* sank at the Crosbie wharf in St John's in January 1949, but was salvaged (*ET,* 5, 18 and 27 January, 4 February 1949). She was finally scuttled at Williamsport when she became unseaworthy (*ET,* 27 February 1957).

98 *ET,* 29 March 1952.

99 The closure of Newfoundland whaling was a matter of "concern to the Government, guarantors to one of the companies of a bank loan of $375,000" *(ET,* 25 April 1952).

100 *ET,* 29 January, 21 August, 19 November, 11 December 1952.

101 *ET,* 5 December 1953; 1 March 1961.

102 *ET,* 4 July 1959.

103 *ET,* 11 January, 27 February, 5 March 1957. The vessel was built in 1916, at 209 tons

gross, length 109.4 feet, depth 13.6 feet, breadth 214 feet (*ET,* 29 January 1962).

104 It was at Hawke Harbour, for example, that Salvesen developed the use of nylon foregoers (the length of rope nearest the harpoon), which then became the industry standard (Vamplew 1975, 231).

CHAPTER TEN

1 *ET,* 27 August 1947. J.C. Ellis and Associates formed the company, with an operating capital of $10,000. Another source relates that Mr Charles Ashwell, a director of the Ontario Fur Breeder's Association, was instrumental in developing the pothead fishery at Dildo, and became managing director of the company (Ashwell, n.d.).

2 *ET,* 22 January 1949.

3 *ET,* 11 May 1946.

4 *ET,* 7 June 1946. Operations were managed by a Norwegian, Ivor Iversen, who "arrived here [St John's] in an open motor boat from North Sydney. The trip was made in 60 hours." Iversen and Victor Clouston were directors, and Eric Martin the plant manager (*ET,* 21 February 1957). Captain Iversen had previously been involved in collecting and selling cod liver oil. Vaughan George (former whaler), pers. com., Dildo, 25 August 2004).

5 *ET,* 30 March 1948. The harpoon gun used was imported from Norway. On one occasion, two whales were reportedly killed with "one shot which went clear through the first" (*ET,* 27 August 1947). Non-explosive harpoons were used to kill potheads (V. George, pers. com., 25 August 2004).

6 *ET,* 15 December 1951. The vessel was built to the design of Captain Iversen (V. George, pers.com., 25 August 2004).

7 26 May 1951. See also 11 December 1953.

8 *ET,* 26 April 1957.

9 *ET,* 30 May, 20 September 1952; 18 November 1953. According to Ashwell (n.d), the first railcar load of whale meat was shipped to Mr Don Gibson, an Ontario fur farmer, on 20 July 1951.

10 *ET,* 10 November 1957.

11 *ET,* 22 November 1954.

12 *ET,* 14 December 1953.

13 *ET,* 21 August 1954. The electrocution of potheads retained within large "booms" was also unrealistically proposed (*ET,* 13 July 1957).

14 *ET,* 8 October 1954.

15 *ET,* 25 October 1954.

16 *EH,* 20 April, 19 July 1955.

17 *ET,* 6 April 1955.

18 *ET,* 2 May, 5 August 1955.

19 It was felt, for example, that "drinking wells near the sea [would be] filled with maggots, and [would] thus [be] unfit for human use" (*ET,* 10 September 1957).

20 These, he claimed, were "selected from a list of literally hundreds of applications which [were] on file" (*ET,* 27 July 1955).

21 *ET,* 19 March 1957.

22 The 1956 pothead catch in Trinity Bay was considered "satisfactory," with 150,000 pounds of meat being exported to Illinois (*EH,* 6 August 1957). This, and the killing of "10,000" animals in July-August 1956 (*ET,* 10 August 1957), undoubtedly had a negative impact on stocks; a "shortage" of potheads was reported in 1958 (*ET,* 29 September 1958).

23 *ET,* 8 January 1958. The fish was purchased from the adjacent fish plant of O'Brien Bros (*ET,* 2 September 1963).
24 Tønnessen and Johnson 1982, 753.
25 "Arctic Fishery Products are at the moment confining themselves to freezing squid and processing whale oil" (*ET,* 5 August 1956). See also 10 May 1957.
26 *ET,* 11 June 1957.
27 *ET,* 31 May 1957.
28 *ET,* 2 January 1958.
29 *ET,* 21 September 1960.
30 *ET,* 4 October 1960. V. George, pers. com., Dildo, 25 August 2004. Captain Clarence George is memorialized by his family as the "first native whaling skipper in Trinity Bay from 1945–1972" (plaque in the Dildo and Area Interpretation Centre, August 2004).
31 *ET,* 13 June 1960. *Eriksen* was seventeen tons gross, and forty-five feet in length.
32 *ET,* 9 September 1959. Some fishermen on the west coast of Newfoundland also used the slipway of the abandoned factory at Lark Harbour for hauling out and flensing carcasses (*ET,* 4 August 1960). In 1957 the meat requirements of local mink ranchers were satisfied, with a limited quantity left for export (*ET,* 10 September 1957).
33 *ET,* 23 August 1962.
34 *ET,* 16 October, 27 November 1963.
35 *ET,* 4 February 1964.
36 *ET,* 24 July 1964. Only 156,000 pounds of whale meat were produced locally in 1963, yet 500,000 pounds was considered necessary to sustain the mink industry.
37 Ibid.
38 *EH,* 5 August 1965.
39 *ET,* 29 June 1967.
40 Nevertheless, fur farming continued on a reduced scale. In 1970, for example, 1,400 mink were imported to ranches near Dildo (*ET,* 10 April 1970).
41 *ET,* 8 March 1968.
42 V. George, pers. com., Dildo, 25 August 2004.
43 *ET,* 12 December 1968.
44 Only 324 potheads were captured by vessels between 1948 and 1971 (IWS).
45 The accumulation of carcasses on the beaches "did not encourage visitors to linger in the strong atmosphere" (*ET,* 2 October 1947).
46 *ET,* 17 September 1955. Before the Dildo plant was built, the oil had been inefficiently rendered by the fishermen and sold to dealers (*ET,* 12 August 1930).
47 The Newfoundland branch of the SPCA recommended that a "more humane" method of killing be used (*ET,* 24 September 1970).
48 *ET,* 20 and 21 July 1939; 10 September 1952.
49 *ET,* 23 September 1950. "Considerable difficulty" was experienced hauling a (fin?) whale up the slip in 1952 (*ET,* 28 June 1952).
50 *ET,* 4 October 1952. The company also processed a killer whale which ran aground at Chapel Arm in June 1955, but a school of twenty-five trapped by pack ice two years later could not be used since they were too large to be moved, and could not be flensed on the beach (*ET,* 17 June 1955; 9 May 1957).
51 *ET,* 8 April 1952.
52 *ET,* 1 June 1957.
53 *ET,* 21 September 1959.
54 Two of the fin whales caught in 1959 were sold to the Fur Farmers Feed Cooperative Society, Trinity Bay. The co-op produced twenty-three and eight barrels of whale oil in

1959 and 1960 respectively, and nineteen barrels of sperm oil in 1960. Twenty-nine tons of meat were also produced.

55 *ET,* 5 October 1959; 1 December 1961.

56 Although intended for scrap, *Southern Foam* was scuttled, and *Sposa* was beached at Conception Harbour.

57 *ET,* 1 and 23 September 1959; 6 October 1960.

58 *ET,* 4 April 1962.

59 For further information on Borgen, see Dickinson and Sanger 1999a.

60 According to *ET,* 10 October 1957, at least $3,000 could be earned for six months work. See also *ET,* 21 January 1957. Many of the first to go that year were from Grand Bank and Fortune on the Burin Peninsula, and "more signed up than were required" (*ET,* 22 March 1957).

61 From $265 to $575 for engineers, $265 to $510 for cooks, and $250 to $490 for fireman and deckhands.

62 *ET,* 30 January, 4 March 1959.

63 R.L. Webb 1988b, 275–6. The Newfoundlanders working for the Western Whaling Co. at Coal Harbor also caused the UFAW to reject a proposal from the company to hunt in 1959 under the 1958 contract.

64 A record 869 whales were processed (*ET,* 9 August 1962). Twelve Newfoundlanders were employed, all from Grand Bank (*ET,* 2 October 1959). In 1959, they earned on average $3,800.

65 *ET,* 27 July, 18 December 1961; 16 April 1962.

66 Tønnessen and Johnson, 1982, 582.

67 *ET,* 27 September 1962.

68 *ET,* 17 October 1962. The employer contributed only $75.

69 The Western Canada Whaling Company relented in 1963, agreeing to pay the one-way air fare to Vancouver (*ET,* 3 and 25 April 1963; 29 March 1966). The Newfoundland contingent also received 5 per cent wage and 5 per cent bonus increases for the 1965 season, giving earnings of $3,000 to $5,000 each for six months work from early April (*ET,* 8 February 1965; 29 March 1966).

70 *ET,* 13 December 1967.

71 *ET,* 16 May 1966. The new company was essentially the Arctic Fishery Products Co. Ltd under another name (V. George, pers. com., Dildo, 25 August 2004).

72 Scaplen 1967. The vessel reportedly sustained damage and did not return (V. George, pers. com., Dildo, 25 August 2004). What might be her hulk is aground near Harbour Buffet (D. Flynn, pers. com., Conception Harbour, 30 August 2004).

73 *Westwhale 8* was built in 1953 at Kaldnes Mekaniske Verksted, Tønsberg. She began whaling as *Suderøy XVII* with the Antarctic pelagic ship *Suderøy* in 1952. After the 1958–59 season, she worked for A/S Kosmos until 1961 when she was sold to Taiyo Gyogyo K.K. and renamed *Toshi Maru 22.* She arrived at Coal Harbour in 1963, and became *Westwhale 8.* Bakka 1992, 191. *Westwhale 8* operated from Dildo until whaling ended in 1972. She was bought by R.W. Wilson of St John's in 1974 and sold for scrap in May 1977 to Dartmouth Salvage, Nova Scotia. Borgen died in 1972 of natural causes when working in the Phillippines. R.L. Webb 1988b, 285. *Westwhale 4*, a sister ship bought by Arctic Fishery Products Co. Ltd, was refitted as a herring purse seiner and sent to Harbour Breton.

74 V. George, pers. com., 25 August 2004, Dildo. Those Newfoundlanders working at Dildo during the last three years of the industry received an amount equivalent to their two years best salary (of the three) from the Canadian government. Those who worked

for two years received their one year highest salary, and those who worked for one year received half their annual salary. Rooney, D., to Clarence George, 4 December 1973.

75 Whales were placed on the endangered species list by the United States Department of the Interior (*ET*, 8 August 1969; 2 March 1971). A small quantity of particularly prized cuts of meat was exported to Japan. *ET*, 22 August 1969.

76 *ET*, 28 November 1969.

77 See, for example, *ET*, 30 May 1967, 13 June 1969; 5 March 1971.

EPILOGUE

1 The first attempt to internationally regulate whaling began in Norway with the establishment of the Bureau of International Whaling Statistics in 1930 (*International Whaling Statistics* 1930–52). The *International Convention for the Regulation of Whaling (ICRW)* 1946 was signed on 2 December and became effective on 10 November 1948. The International Whaling Commission (IWC) was established from the *Convention*. In 2003 it included fifty-one member states.

2 See, for example, K.R. Allen 1970; J.E. Gulland and Boerema 1973; Mitchell 1974b. The *Kyo Maru 7* was chartered in 1966 to spend three months tagging whales in the western North Atlantic (*ET*, 16 May 1966).

3 A survey in 1980 provided minimum abundance estimates (Hay 1982).

4 The IWC decided in 1989 "that because of the considerable scientific uncertainty over the numbers of whales of different species and in different geographical stocks, that it would be better not to give population figures except for those species/stocks which have been assessed in detail, and for which there is statistical certainty with respect to the numbers" (http://www.iwcoffice.org/Estimate.htm). The annual NFMS estimates provide the best perspective. See, for example, Waring et al. 2001.

5 See, for example, Campbell 2001. An ongoing update is provided by the COSEWIC web site.

6 *ET*, 30 November 1972.

7 See, for example, K.R. Allen 1973; Mitchell 1972b and 1974b.

8 Quotas are variously recorded. See, for example, *Report of the International Whaling Commission 23*, 1973.

9 An amendment of 14 October 2002 to the ICRW 1946 details current conservation and management policies.

10 For fuller details, see Meredith and Campbell 1988; Sergeant 1977; Waring et al. 2001.

11 The IWC designates three stock categories. Protection Stocks (PS) are protected from commercial whaling. Initial Management Stocks (IMS) and Sustained Management Stocks (SMS) can be commercially hunted under certain conditions (ICRW 1946, amended 14 October 2002).

12 See, for example, Campbell 2001; Mansfield 1985; Waring et al. 2001.

13 Mitchell 1973. Nineteen were taken from Dildo, fifteen from Williamsport, and seven from Blandford.

14 For fuller details, see Campbell 2001; Hay 1985; Mitchell and Reeves 1983; Whitehead 1982 and 1987.

15 Campbell 2001; Nelson and Lien 1996; Waring et al. 2001.

16 Perhaps due to lack of squid, their main food (J. Lien, Whale Research Group, MUN, St John's, pers. com., 2003).

17 See, for example, Campbell 2001; Mitchell 1972a and 1975b. Waring et al. (2001) state that the lack of recent data prohibits population estimates.

18 ICRW 1946 (amended 14 October 2002); Waring et al. 2001.

19 The IWC recognizes four stocks of minkes in the North Atlantic (Campbell 2001; Donovan 1991; Waring et al. 2001).

20 Conference on the Human Environment, 1972.

21 *ET,* 27 July 1972. See also *Marine Mammal Protection Act* 1972. Indigenous peoples in Alaska and on the coasts of the north Pacific and Arctic oceans are exempt for reasons of subsistence and the production of clothing and handicrafts.

22 Whales are protected under Appendix 1 of the *Convention on International Trade in Endangered Species of Wild Fauna and Flora* 1973. CITES was drafted following a resolution adopted by the IUCN. It became effective on 1 July 1975.

23 For a comprehensive assessment of the controversy that ensued, see Holt 2002.

24 *Cetacean Protection Regulations* (1982). The minister may issue hunting licences ($5 per whale other than beluga, narwhal, or bowhead, which are $25) but is highly unlikely to do so. These regulations were replaced in 1993 by the *Marine Mammal Regulations.*

25 For details of global commercial and indigenous whaling, see "Report of the Fifty-Sixth Annual Meeting," IWC 2003. Cole (2003) provides a comprehensive if slightly "environmentally" biased account of the pro-whaling and anti-whaling "debate."

26 Altherr 2003 and Petursddottir 1997 are representative of the literature on Icelandic whaling.

27 Information is widely scattered. This paragraph draws heavily from "Catch Statistics for Marine Mammals in the North Atlantic" 2003; INWRD (no. 17, June 1999; no. 22, December 2001; no. 23, July 2002; no. 24, December 2002; no. 25, July 2003); "Report Status of Marine Mammals in the North Atlantic: The Minke Whale" n.d.; "Statistics of the Government of the Faeroe Islands" 2003.

28 INWRD no. 20, December 2000; no. 23, July 2002.

29 "Canada Withdraws from the International Whaling Convention and Commission" 1981.

30 "Agreement on Cooperation, Research, Conservation and Management of Marine Mammals in the North Atlantic" 1992.

31 Lien 2001.

BIBLIOGRAPHY

PRIMARY DOCUMENTS

Act to Provide for the Formation of a Fisheries Commission and for Other Purposes. 1889. St John's: Government of Newfoundland.

Act to Regulate the Whaling Industry. 1902. *Acts of the General Assembly of Newfoundland Passed in the 2nd Year of the Reign of His Majesty King Edward VII.* St John's: Government of Newfoundland.

Acts of the General Assembly of Newfoundland: Passed in the 2nd Year of the Reign of His Majesty King Edward VII. 1902. St John's: Government of Newfoundland.

"Agreement on Cooperation, Research, Conservation and Management of Marine Mammals in the North Atlantic." 1992. NAMMCO, 9 April. http://www.nammco.no.

Aquaforte Whaling Co. Contract Agreement, 16 November 1901. Ellefsen Papers (hereafter EP).

– 13 May 1902. EP.

– 7 June 1902. EP.

– Contracts. n.d. EP.

– Employment and Wage Ledger, 1902–03. EP.

– Letter Book, 21 April 1905. EP.

– Letter Book, 18 January 1906. EP.

– Letter Book, 12 September 1907. EP.

– Letter Book, 14 May 1908. EP.

– Shareholder Agreement, 30 October 1901. EP.

– Whale and Oil Ledger. 1902–08. EP.

Ashwell, T. n.d. Documents, Whaling and Sealing Museum, South Dildo, Nfld.

Atlantic Whaling and Manufacturing Co. Ltd. 1904. St John's: Registry of Companies, drawer 1, no. 40.

– 1912. St John's: Registry of Companies, no. 208, paper no. 1737.

Barkham, J.S. 1981. "The Basque Whaling Establishment in Labrador: 1536–1603." Unpublished paper presented at the Conference on the Discovery of the Arctic Region As Seen Through the Descriptions of Travellers and the Work of Cartographers from Early Antiquity to the Eighteenth Century. Rome, October.

British Norwegian Whaling Co. Ltd. 1926–34. St John's: Registry of Companies, drawer 9, no. 9.

Cabot Steam Whaling Co. Ltd. 1896–1914. St John's: Registry of Companies, drawer 1, no. 12, box 47.

"Canada Withdraws from the International Whaling Convention and Commission." 1981. Ottawa: Department of External Affairs, Communique. Reprinted in *INWRD* no. 20, December 2000.

Cape Broyle Whaling and Trading Co. Ltd. 1902–06. St John's: Registry of Companies, drawer 1, no. 55.

"Catch Statistics for Marine Mammals in the North Atlantic." 2003. Lofoten, Norway. http://www.highnorth.no.

"Cetacean Protection Regulations Introduced into the Fisheries Act." 1982. *Canada Gazette*, part 2 (SOR/82–614): 2188.

Christian Salvesen Co. Ltd, Hawke Harbour Operations. n.d. Salvesen Papers (hereafter SP).

– Letterbook. n.d. SP.

– List of the Archives of Messrs Christian Salvesen Limited. n.d. SP.

– Notes, 8 October 1939. SP.

– Notes, 27 May 1943. SP.

– Notes, 2 August 1943. SP.

– Position of Ships Book. n.d. SP: D/13–14.

– Whaling Officers. n.d. SP: G/12.

Conference on the Human Environment. 1972. Stockholm: UN Resolution 33.

Convention on International Trade in Endangered Species of Wild Fauna and Flora, Appendix 1. 1973. Washington, D.C.: *MMPA*.

COSEWIC web site. http://www.cosewic.gc.ca.

Crew List, *Sobraon* 1907. St John's: Maritime History Archive, MUN.

Harbour Grace Whaling Co. Ltd. 1904. St John's: Registry of Companies, drawer 2, no. 91.

Harris, L. 1987. "The Outport Phenomenon: Life at the Edge of the Sea." Typescript of paper presented to the Summit of the Sea Conference, St John's: 1–25.

Hercules. 1831. Personal journal maintained by the medical officer of vessel *Hercules*. Aberdeen: University Library, King's College.

International Convention for the Regulation of Whaling. 1946. Washington, D.C.

International Whaling Statistics. 1930–52. Oslo: Bureau of International Whaling Statistics.

– 1930–84. Oslo: Committee for Whaling Statistics, *Norske Hvalråds Statistike Publikasjoner*.

In the Matter of the Companies Act and in the Matter of the Newfoundland Whaling Company Limited. n.d. Attachment to Liquidator's Report, Supreme Court of Newfoundland, no. 132, 1938. SP

IWC. 1989. http://www.iwcoffice.org/Estimate.htm.

Jas. Murphy and Sons. 1916. Letterhead. Whaling Box. Vancouver Maritime Museum (hereafter VMM).

Job, W.C. 1903. "Whaling Notes Compiled by William C. Job on a Visit to Norway." Typescript. Job Papers (hereafter JP). St John's: PANL.

Job Bros. Memorandum of Agreement, 13 February 1904. JP, no. 6.

Kosmos, A/S, and Anders Ellefsen, 18 August 1911: Agreement. EP.

Labrador Whaling and Manufacturing Co. Ltd. 1904. St John's: Registry of Companies, drawer 2, no. 89.

Le Marquand, A. n.d. Taped interview with J. Goddard. BCPA: 3945 7–9.

MacPherson, D.B. 1985. Typescript, n.p.: BCPA.

Marine Mammal Protection Act. 1972. Washington, D.C.: U.S. Congress, USC 1361.
Marine Mammal Regulations. 1993. SOR/93–56.
Marmo, N. n.d. Taped interview with J. Goddard. BCPA: 3945.
Mic Mac Whaling Co. Ltd. 1903. St John's: Registry of Companies, drawer 1, no. 7.
Newfoundland. Minutes of the Executive Council, 13 September 1899. PRO/CO 126.
– 24 March 1902. PRO/CO 194.
Newfoundland and Labrador Heritage Website. n.d. St John's: Sponsored by Memorial University of Newfoundland, the J.R. Smallwood Foundation, and the CRB Foundation. http://www.heritage.nf.ca.
Newfoundland Department of Health. Death certificate of Adolph Nielsen. 1903. St John's: Government of Newfoundland.
Newfoundland Steam Whaling Co. Ltd. 1900–10. St John's: Registry of Companies, box 41, no. 37.
Newfoundland Whaling Co. Ltd. n.d. SP.
– 1923–36. St John's: Registry of Companies, drawer 8, no. 623.
– Directors' Minutes, 16 January 1934. SP.
– Directors' Minutes, 9 March 1938. SP.
– Directors' Minutes, 1 April 1938. SP.
– Directors' Minutes, 6 April 1938. SP.
– Directors' Minutes, 10 May 1938. SP.
North Pole. 1837. Logbook of *North Pole.* Dundee Public Library.
Notary Public Document, 6 January 1905. UBC: AW 15474 V. 41.
Olsen Whaling and Sealing Co. Ltd. 1940–45. St John's: Registry of Companies, no. 1667.
Pacific Whaling Co. Ltd. 1909. BCPA: AD MSS. 21, V. 2, fol. 1–4.
Polar Whaling Co. Ltd. Ledger. 1930–47. SP: A/24.
– New Account. 1930-47. SP: A/24.
"Report of the Fifty-sixth Annual Meeting." 2003. Berlin: IWC, Scientific Committee.
Report of the Fisheries Commission Appointed by His Excellency the Governor-in-Council to Investigate the Operations of Fisheries Departments in Other Countries. 1888. St John's: Government Publication.
Report of the International Whaling Commission. 1973. No. 23. Cambridge, U.K.: IWC.
Report of the Interdepartmental Committee on Research and Development in the Dependencies of the Falkland Islands. 1920. London: His Majesty's Stationary Office cmnd. 657.
"Report Status of Marine Mammals in the North Atlantic: The Minke Whale." n.d. Tromsø, Norway: North Atlantic Marine Mammal Commission. http://www.nammco.no.
Revised Statutes of Canada. Ch 45, 4 Edw. 7, Ch. 13.
Rose-au-Rue Whaling Co. Ltd. 1910–15. St John's: Registry of Companies, drawer 3, no. 181.
St Mary's Whaling and Trading Co. Ltd. 1903. St John's: Registry of Companies, drawer 1, no. 71.
Salvesen, H.K. 1935. Summary Draft, Matters of Finance between Messrs Chr. Salvesen and Co., and Newfoundland Whaling Co., and/or Capt. Anonsen the Late Manager. April. SP.
– n.d. Summary of Notes Taken during Conversation with Capt. Anonsen. SP.
"Statistics of the Government of the Faeroe Islands." 2003. http://www.whaling.fo.
Stevenson, R.L. 1950–51. Diary, Hawke Harbour Whaling Station, 1950–51, by R.L. Stevenson, inspector, Canadian Department of Fisheries. NSA.
"Summary of Costs for Getting the Sulphurbottom Skeleton for St Louis." n.d. Washington, D.C.: Smithsonian Institution Archives, record unit 7181, box 2.
"Summary Table of Eastern Canada Catches As Coded by IWC, 1946–72." 1991. Cambridge, U.K.: IWC.

Trinity Whaling and Manufacturing Co. Ltd. 1904. St John's: Registry of Companies, drawer 2, no. 90.
True, F.W. 1899. Personal notebooks. Washington, D.C.: Smithsonian Institution.
Victoria Whaling Co. Ltd. 1916. Letterhead. VMM, box 2.

CORRESPONDENCE

Admiralty to Salvesen, 29 August 1940. SP.
– 11 June 1943. SP.
– 13 April 1944. SP.
A.H. Murray and Co. Ltd. to Gorton Pew, 30 May 1941. SP.
A.H. Murray and Co. Ltd. to Mardon-Wild, 27 May 1941. SP.
A.H. Murray and Co. Ltd. to Salvesen, 24 September 1940. SP.
– 23 October 1940. SP.
Anonsen, A., to Salvesen, 31 December 1936. SP.
– 1 February 1937. SP.
– 23 July 1937. SP.
Anonsen, Amund, to Chr. Salvesen Co. Ltd, 31 December 1936. SP.
Anonsen vs. the Newfoundland Whaling Company, 1941. SP.
Atlas Insurance Co. to A.H. Murray and Co. Ltd, 28 December 1939. SP.
Balcom, S., to R.P. Venning, 2 September 1907. UBC: AW 15474 vol. 41.
Berg, R., to Anders Ellefsen, 17 July 1906. EP.
Brister and Son to Anders Ellefsen, 25 June 1908. EP.
Bryde, Johan, to Anders Ellefsen, 8 November 1906. EP.
Carter, John, to Anders Ellefsen, 7 March 1906. EP.
Clift, Thomas B., to Andreas Ellefsen, 5 December 1906. EP.
Colonial Cordage Co. to Anders Ellefsen, 19 March 1906. EP.
Colonial Secretary to C. O'Neill Conroy, 11 October 1901. EP.
Conroy, C. O'Neill, to Anders Ellefsen, 18 February 1903. EP.
– 5 September 1903. EP.
– 10 October 1903. EP.
Conroy, C. O'Neill, to T.J. Murphy, 18 November 1902. PANL.
Cook, Tasker, to H.M. Maddick, 4 January 1930. St John's: Registry of Companies, no. 200.
Ellefsen, Anders, to Andreas Ellefsen, 21 April 1905. EP.
– 10 October 1908. EP.
Ellefsen, Anders, to Colonial Cordage Co., 7 August 1906. EP.
Ellefsen, Anders, to Dr L. Rissmüller, 4 May 1903. EP.
Ellefsen, Anders, to Harvey and Company, 1 August 1908. EP.
Ellefsen, Anders, to Job Brothers, 25 August 1906. EP.
– 2 April 1908. EP.
Ellefsen, Anders, to Joseph Salters, North Sydney, N.S., 9 July 1908. EP.
Ellefsen, Anders, to Joseph Salters, 11 July 1908. EP.
Ellefsen, Anders, to J.R. Hunter, 24 July 1903. EP.
Ellefsen, Anders, to Messrs F.W. Bissett and Co., 27 June 1908. EP.
Ellefsen, Anders, to P.F. Windsorton, 25 July 1908. EP.
Ellefsen, Anders, to R. Berg, 21 April 1905. EP.
Ellefsen, Anders, to T. Clift, 24 November 1906. EP.
Ellefsen, Andreas, to Anders Ellefsen, 2 October 1906. EP.
– 2 November 1906. EP.
– 12 December 1906. EP.

– 20 October 1908. EP.
Ellefsen, Arne, to Anders Ellefsen, 22 March 1907. EP.
Finn, R.E., to J-R.F. Préfontaine, 11 July 1904. UBC: AW 15474 V. 41.
– 29 July 1904. UBC: AW 15474 V. 41.
Fox, Knight and Phelan to Salvesen, 25 January 1940. SP.
Gourdeau, F., to J. Pope, 26 November 1903. UBC: AW 15474 V. 41.
Sorvald, Håkon, to C.W. Sanger, 22 December 1992. Possession of C.W. Sanger.
Huffman Manufacturing Co. to Anders Ellefsen, 13 August 1903. EP.
Huffman, W.F., to Anders Ellefsen, 29 July 1903. EP.
– 23 October 1903. EP.
Hunter, J.R., to Anders Ellefsen, 26 May 1903. EP.
– 3 June 1903. EP.
– 16 June 1903. EP.
– 30 June 1903. EP.
– 16 July 1903. EP.
Jas. Murphy and Sons to Victoria Whaling Co. Ltd, 17 May 1916. VMM: Whaling Box.
– 11 September 1916. VMM: Whaling Box.
Job Brothers to Anders Ellefsen, 23 August 1906. EP.
– 27 September 1906. EP.
Job Brothers to Andreas Ellefsen, 27 May 1903. EP.
Maclellan, J., to Anders Ellefsen, 6 July 1906. EP.
Ministry of Shipping to Salvesen, 7 February 1940. SP.
– 15 March 1940. SP.
Munroe, J.H., to Anders Ellefsen, 10 August 1906. EP.
Northfield to Ellefsen, 6 December 1907. EP.
Olsen, O., to Salvesen, 28 May 1940. SP.
Payne, W.B., to Anders Ellefsen, 11 September 1906. EP.
Polar Whaling Co. Ltd to General Registrar and Record Office of Shipping and Seamen, London, 24 June 1943. SP.
Préfontaine, J-R.F., to A. Johnson, 14 November 1903. UBC: AW 15474 V. 41.
Préfontaine, J-R.F., to A. Walker, 14 November 1904. UBC: AW 15474 V. 41.
Rankin, C.B., to Anders Ellefsen, 5 December 1906. EP.
– 11 December 1907. EP.
– 21 September 1908. EP.
Rissmüller, L., to Anders Ellefsen, 5 June 1903. EP.
– 16 June 1903. EP.
Robert Shankland and Co. to Anders Ellefsen, 20 December 1906. EP.
Rooney, D., to Clarence George, 4 December 1973. Whaling and Sealing Museum, South Dildo, Nfld.
Ruck, S.D., to Department of Immigration [Victoria], 6 April 1916. VMM: Whaling Box.
Salvesen to A. Anonsen, 29 October 1936. SP.
– 16 April 1937. SP.
Salvesen to Admiralty, London, 8 January 1943. SP.
– 27 May 1943. SP.
– 19 June 1943. SP.
– 15 February 1944. SP.
Salvesen to A.H. Murray Co. Ltd, 15 November 1939. SP.
Salvesen to Board of Trade, London, 26 May 1937. SP.
– 13 October 1939. SP.
Salvesen to Food (Defense Plans) Department, London, 28 September 1938. SP.

Salvesen to Liquidator, Newfoundland Whaling Co. Ltd, 27 February 1940. SP.
Salvesen to Ministry of Shipping, London, 2 February 1940. SP.
– 8 February 1940. SP.
– 10 February 1940. SP.
– 29 February 1940. SP.
– 11 April 1940. SP.
– 6 August 1940. SP.
– 4 November 1940. SP.
Salvesen to Newfoundland Co. Ltd, 1 February 1937. SP.
Salvesen to Post and Telegraph Censorship Department, London, 13 May 1943. SP.
Trading with the Enemy Department to Salvesen, 20 July 1942. SP.
Victoria Whaling Co. Ltd to Jas. Murphy and Sons, 21 September 1916. VMM: Whaling Box.
Way, C., to Anders Ellefsen, 3 September 1903. EP.
– 19 September 1903. EP.

NEWSPAPERS AND JOURNALS

Newspapers

Compass [*TC*], Conception Bay Centre, Nfld.
Daily News [*DN*], St John's, Nfld.
Dundee Advertiser [*DA*], Dundee, Scotland.
Evening Chronicle [*EC*], St John's, Nfld.
Evening Herald [*EH*], St John's, Nfld.
Evening Mercury [*EM*], St John's, Nfld.
Evening Telegram [*ET*], St John's, Nfld.
Globe and Mail [*GM*], Toronto, Ont.
Morning Chronicle [*TMC*], St John's, Nfld.
New York Fishing Gazette [*NYFG*], N.Y.
Pacific Fisherman [*PF*], Portland, Oreg.
Peterhead Sentinel [*PS*], Peterhead, Scotland.
Royal Gazette [*RG*], St John's, Nfld.
Telegram [*TT*], formerly the *Evening Telegram*, St John's, Nfld.
Victoria Colonist [*VC*], Victoria, B.C.
Western Star [*WS*], Corner Brook, Nfld.

Journals

Hvalfangerforeningen. 1927–30. Oslo: Association of Whaling Companies.
Hvalfangernes Assuranceforening Gjensidig, 1911–36. 1936. Sandefjord: Whalers Mutual Insurance Association.
Norsk Hvalfangst-Tidende. 1929 and various. Oslo, Norway.

SECONDARY SOURCES

"Aboriginal Peoples." n.d. NLHWS.
Abstract Census and Return of the Population, &c. of Newfoundland, 1857. 1858. St John's.
Agirreazkuenaga Zigorraga, J., R. Basurto Larranaga, R. Lopez Atxurra, and R. Mieza Mieg. 1980. Historia de Euskal Harria. San Sebastean: Ediciones Vascas.
Aldrich, F.A. 1988. "Harvey, Moses." In *Encyclopedia of Newfoundland and Labrador* vol. 2, 849–50.

Alexander, D. 1974. "Development and Dependence in Newfoundland." *Acadiensis* 4, no. 1: 3–31.

– 1976. "Newfoundland's Traditional Economy and Development to 1934." *Acadiensis* 5, no. 2: 56–78.

– 1977. *The Decay of Trade: An Economic History of the Newfoundland Saltfish Trade, 1935–1965*. St John's: Institute of Social and Economic Research, Memorial University of Newfoundland.

– 1980. "Newfoundland's Traditional Economy and Development to 1934." In *Newfoundland in the Nineteenth and Twentieth Countries: Essays in Interpretation*, edited by J.K. Hiller and P. Neary, 17–39. Toronto: University of Toronto Press.

– 1983. *Atlantic Canada and Confederation: Essays in Canadian Political Economy*, compiled by E.W. Sager, L.R. Fischer, and S.O. Pierson. Toronto: University of Toronto Press.

Allen, G.M. 1928. "Whales and Whaling in New England." *Scientific Monthly* 27: 340–3.

Allen, J.A. 1908. "The North Atlantic Right Whale and Its Near Allies." *Bulletin of the American Museum of Natural History* 24: 277–330.

Allen, K.R. 1970. "A Note on Baleen Stocks of the North-West Atlantic." *Report of the International Whaling Commission* 20: 112–13.

– 1973. "Catch Per Unit Effort of Northwest Atlantic Fin Whale Stocks." *Report of the International Whaling Commission* 23: 97–8.

– 1980. *Conservation and Management of Whales*. Seattle: University of Washington Press.

Altherr, S. 2003. *Iceland's Whaling Comeback: Preparation for the Resumption of Whaling*. Munich: Pro Wildlife.

Andrews, R.C. 1929. *Ends of the Earth*. New York: Putnam.

Annual Report of the Newfoundland Fisheries Commission for the Year 1889. 1890. St John's: Government Publication.

Anspach, L.A. 1819. *A History of the Island of Newfoundland: Containing a Description of the Island, the Banks, the Fisheries, and Trade of Newfoundland and the Coast of Labrador*. London: T. and J. Allman.

Antler, S.D. 1975. "Colonial Exploitation and Economic Stagnation in Nineteenth Century Newfoundland." PhD thesis, University of Connecticut.

Arlov, T.B. 1993. "The Role of Whaling in the Struggle for Supremacy over Svalbard (Spitsbergen)." In *Whaling and History: Perspectives on the Evolution of the Industry*, edited by B.L. Basberg et al., 81–90, Sandefjord: Chr. Christensens Hvalfangstmuseum.

"Atlantic Realm, The." 1987. *In Historical Atlas of Canada*, vol. 1, edited by R.C. Harris and G.J. Matthews, 47–51. Toronto: University of Toronto Press.

Baker, M. 1988. "Government." In *ENL*, vol. 2, 580–661.

– 1993. "Political Parties." In *ENL*, vol. 4, 357–63.

Baker, M., A.B. Dickinson, and C.W. Sanger. 1992. "Adolph Nielsen: Norwegian Influence on Newfoundland Fisheries in the Late 19th – Early 20th Centuries." *Newfoundland Quarterly* 27, no. 2: 25–35.

Bakka, Jr, D. 1992. *Hvalfangsten, Eventyret tar Slutt*. Larvik, Norway: Kronhn Johansien.

Barkham, S. 1974. "The Spanish Province of Terra Nova." *Canadian Archivist* 2, no. 5: 73–83.

– 1976. "Two Documents Written in Labrador, 1572 and 1577." *Canadian Historical Review* 57, no. 2: 235–8.

– 1977a. "First Will and Testament on the Labrador Coast." *Geographical Magazine* 49, no. 9: 574–81.

– 1977b. "The Identification of Labrador Ports in Spanish Sixteenth Century Documents." *Canadian Cartographer* 14, no. 1: 1–9.

– 1978. "The Basques: Filling a Gap in Our History between Jacques Cartier and Champlain." *Canadian Geographical Journal* 96, no. 1: 8–19.

– 1980. “A Note on the Strait of Belle Isle during the Period of Basque Contact with Indians and Inuit.” *Etudes Inuit Studies* 4, nos. 1–2: 51-8.

– 1982. “Documentary Evidence for sixteenth Century Basque Whaling Ships in the Strait of Belle Isle.” In *Early European Settlement and Exploitation in Atlantic Canada: Selected Papers*, edited by G.M. Storey, 53–95. St John's: Memorial University of Newfoundland.

Basberg, B.L. 1993. “Survival against All Odds? Shore Station Whaling at South Georgia in the Pelagic Era, 1925–1960.” In *Whaling and History: Perspectives on the Evolution of the Industry*, edited by B.L. Basberg et al., 157–67. Sandefjord: Chr. Christensens Havalfangstmuseum.

Basberg, B.L., J.E. Ringstad, and E. Waxelsen, eds. 1993. *Whaling and History: Perspectives on the Evolution of the Industry*. Sandefjord: Chr. Christensens Hvalfangstmuseum.

Bates, A.C. 1994. “Truck System.” In *ENL*, vol. 5, 428.

“Before Cabot.” 1997. NLHWS.

Biggar, H.P., ed. 1911. *The Precursors of Jacques Cartier, 1497–1534*. Ottawa: Publication of the Canadian Archives, no. 5.

Black, W.A. 1960. “The Labrador Floater Fishery.” *Annals of the Association of American Geographers* 50, no. 3: 267–93.

Blake, R. 1993. “The Problem of Newfoundland: The Fisheries and Newfoundland's Integration into Canada, 1948–57.” In *Twentieth-Century Newfoundland: Explorations*, edited by J. Hiller and P. Neary, 239–72. St John's: Breakwater Press.

Bobe, L. 1952. *Hans Egede: Colonizer and Missionary of Greenland*. Copenhagen: Rosenkilde and Bagger.

Bockstoce, J. 1977. *Steam Whaling in the Western Arctic*. New Bedford, Mass.: New Bedford Whaling Museum.

– 1978. “History of Commercial Whaling in Arctic Alaska.” *Alaska Geographic* 5, no. 4: 1–4.

Bonnycastle, R. 1842. *Newfoundland in 1842*. London: Henry Colburn.

Boorstin, D.J. 1983. *The Discoverers*. New York: Random House.

Brebner, J.B. 1933. *The Explorers of North America, 1492–1806*. London: A. and C. Black.

Brière, J.-F. 1997. “The French Fishery in the Eighteenth Century.” In *How Deep Is the Ocean? Historical Essays on Canada's Atlantic Fishery*, edited by J.E. Candow and C. Corbin, 15–30. Sydney, N.S.: University College of Cape Breton Press.

Brown, C. 1972. *Death on the Ice: The Great Newfoundland Sealing Disaster of 1914*. Toronto: Doubleday Canada.

Brown, P.W. 1909. *Where the Fishers Go: The Story of Labrador*. Toronto: Musson Book Co.

Brown, S.C. 1993. “Norse Discovery.” In *ENL*, vol. 4, 100–4.

Bruce, W.S. 1907. “Reports on the Scientific Results of the Voyage of the S.S. *Scotia*.” In *Scottish National Antarctic Expedition, 1902–04*. Edinburgh: Scottish Oceanographic Laboratory.

Bruun, D. 1918. “The Icelandic Colonization of Greenland and the Finding of Vineland.” *Meddelelser om Grønland*, vol. 57.

Budden, G.E. 1988. “French Shore.” In *ENL*, vol. 2, 407–15.

Budker, P. 1958. Whales and Whaling. London: George G. Harrap.

Burke, N.R. 1971. “Some Observations on the Migration of Labourers from the South of Ireland to Newfoundland in Pre-Famine Times.” *Journal of the Cork Historical and Archaeological Society* 76: 95–109.

Bursey, B.C. 1993. “Petroleum Exploration and Development.” In *ENL*, vol. 4, 267–71.

Butt, J. 1998. “The Decline of the Migratory Fishery.” NLHWS.

Cadigan, S. 1995. *Hope and Deception in Conception Bay: Merchant-Settler Relations in Newfoundland, 1785–1855*. Toronto: University of Toronto Press.

Campbell, R.R. 2001. “Rare and Endangered Fishes and Marine Mammals of Canada:

COSEWIC Fish and Marine Mammals Subcommittee. Status Report XIV." *Canadian Field-Naturalist* 115, no. 4: 564–72.

Carson, W. 1813. *Reasons for Colonizing the Island of Newfoundland.* Greenock, Scotland: William Scott.

Carus-Wilson, E.M. 1962. *The Merchant Adventurers of Bristol in the Fifteenth Century.* Bristol: Bristol Branch of the Historical Association.

Cashin, P. 1976. *My Life and Times, 1880–1919.* St John's: Breakwater Books.

Cell, G.T. 1969. *English Enterprise in Newfoundland, 1577–1600.* Toronto: University of Toronto Press.

– 1982a. *Newfoundland Discovered: English Attempts at Colonization, 1610–1630.* London: Hakluyt Society.

– 1982b. "The Cupids Cove Settlement: A Case Study of the Problems of Early Colonization. In *Early European Settlement and Exploration in Atlantic Canada: Selected Papers,* edited by G.M. Story. St John's: Memorial University of Newfoundland.

Census of Newfoundland and Labrador, 1891. 1893. St John's: J.W. Withers.

Census of Newfoundland and Labrador, 1911. 1914. St John's: J.W. Withers.

"Christensen, C.L., in Memoriam." 1951. *Norsk-Hvalfangste Tidende* (Year Book) 1950: 7–9.

Christensen, T. 1975. "Preliminary Report on the Norwegian Fishery for Small Whales: Expansion of Norwegian Whaling to Arctic and Northwest Atlantic Waters and Norwegian Investigation of the Biology of Small Whales." *Journal of the Fisheries Research Board of Canada* 32, no. 7: 1083–94.

Ciriquiain-Gaiztarro, M. 1961. *Los vascos en la pesca de la ballena.* San Sebastian: Biblioteca Vascongada de Los Amigos del Pais.

Clark, A.H. 1887. "History and Present Condition of the Fishery." In *The Fisheries and Fishing Industries of the United States,* vols. 1–4, edited by G.B. Goode, 3–218. Washington, D.C.: Government Printing Office.

Cole, B. 2003. "Whaling – A Bloody War." *The Economist,* 30 December.

Colwell, M. 1970. *Whaling Around Australia.* London: Angus and Robertson.

Candow, J.E., and C. Corbin, eds. 1997. *How Deep Is the Ocean? Historical Essays on Canada's Atlantic Fishery.* Sydney, N.S.: University College of Cape Breton Press.

Conway, M. 1900. "Some Unpublished Spitsbergen MSS." *Geographical Journal,* no. 6: 628–36.

– 1904. *Early Dutch and English Voyages to Spitsbergen in the Seventeenth Century.* London: Hakluyt Society.

– 1906. *No Man's Land: A History of Spitsbergen from Its Discovery in 1596 to the Beginning of the Scientific Exploration of the Country.* Cambridge: University Press.

Corish, P.J. 1987. "The Irish Catholics at the End of the Penal Era." In *Religion and Identity: The Experience of Irish and Scottish Catholics in Atlantic Canada,* edited by T. Murphy and C.J. Byrne, 1–17. St John's: Jesperson Press.

Crawley, J.E. 1989. "Empire versus Truck: The Official Interpretation of Debt and Labour in the Eighteenth-Century Newfoundland Fishery." *Canadian Historical Review* 70, no. 3: 311–36.

Crone, G. 1969. *The Discovery of America.* London: Hamilton.

Cuff, R.H. 1991. "Labrador Fishery." In *ENL,* vol. 3, 225–30.

– 2001. "Railway, 1882–1997." NLHWS.

Cuff, R.H., C.F. Hogan, K. Matthews, and S. Ryan. 1988. "Fisheries." In *ENL,* vol. 2, 130–66.

Cumming, W.P., R.A. Skelton, and D.B. Quinn. 1971. *The Discovery of North America.* London: Elek.

Dalgard, S. 1962. *Dansk-Norsk Hvalfangst, 1615–1660.* Kobenhavn: G.E.C. Gad.

Davis, L.E., R.E. Gallman, and K. Gleiter. 1997. *In Pursuit of Leviathan: Technology, Institu-*

tions, Productivity, and Profits in American Whaling, 1816–1906. Chicago: University of Chicago Press.

Davis, R. 1973. *The Rise of the Atlantic Economies.* Ithaca, N.Y.: Cornell University Press.

De Cortazar, J.A. 1966. *Vizcaya en el siglo-XV.* Bilbao: Ediciones de la Caja de Ahorros Vizcaina.

"Decline of the Migratory Fishery." n.d. NLHWS.

De Jong, C. 1978. *A Short History of Old Dutch Whaling.* Pretoria: University of South Africa.

De Veer, G. 1853. *The Three Voyages of William Barents to the Arctic Regions (1594, 1595, and 1596).* London: Hakluyt Society.

Dickinson, A.B. 1989. "Southern Hemisphere Fur Sealing from Atlantic Canada." *American Neptune* 90, no. 4: 278–90.

Dickinson, A.B., and C.W. Sanger. 1990. "Modern Shore-Based Whaling in Newfoundland and Labrador: Expansion and Consolidation, 1898–02." *International Journal of Maritime History* 2, no. 1: 83–116.

– 1991. "A Newfoundland Floating Factory Whaling Expedition." *Polar Record* 27, no. 161: 125–8.

– 1993a. "Modern Shore-Station Whaling in Newfoundland and Labrador: The Peak Season." *International Journal of Maritime History* 5, no. 1: 127–54.

– 1993b. "The Origin and Development of North American Modern Shore Station Whaling: Newfoundland and the Norwegians, 1896–1916." In *Whaling and History: Perspectives on the Evolution of the Industry*, edited by B.L. Basberg, J.E. Ringstad, and E. Wexelsen, 91–100. Sandefjord: Chr. Christensens Hvalfangstmuseum.

– 1999a. "Newfoundland and Labrador Shore-Station Whaling: The Final Demise, 1951–1972." *The Northern Mariner/Le Marin du Nord* 9, no. 3: 39–52.

– 1999b. "Newfoundland and Labrador Shore-Station Whaling: The Third Major Phase, 1937–1951." *International Journal of Maritime History* 11, no. 1: 101–16.

– 2001. *Norwegian Whaling in Newfoundland: The Aquaforte Station and the Ellefsen Family, 1902–1908.* St John's: International Maritime Economic History Association.

Dietz, J. 1923. *Master Johann Dietz: Surgeon in the Army of the Great Elector and Barber to the Royal Court.* Translated by B. Miall. London: George Allen and Unwin.

Donovan, G.P. 1991. "A Review of the IWC Stock Boundaries." *Report of the International Whaling Commission* (special issue) 13: 39–68.

Douglas, W.A., and J. Bilbao. 1975. *Amerikanuak: Basques in the New World.* Reno, Nevada: University of Nevada Press.

Duncan, N. 1904. "The Chase of the Fin-Back Whale." *Outing* 44, no. 6, 679–85.

Economic Council of Canada. 1980. *Newfoundland: From Dependency to Self-Reliance.* Hull, Quebec: Minister of Supply and Services.

"Economy." n.d. NLHWS.

Edmondston, A. 1809. *A View of the Ancient and Present State of the Zetland Islands; Including Their Civil, Political, and Natural History; Antiquities; and An Account of Their Agriculture, Fisheries, Commerce, and the State of Society and Mammals.* Vols. 1–2. Edinburgh: John Ballantyne and Co.

Edmundson, G. 1911. *Anglo-Dutch Rivalry during the First Half of the Seventeenth Century.* Oxford: The Clarenden Press.

"Effects of War on Early Settlement." n.d. NLHWS.

Egede, H. 1818. *A Description of Greenland* 2nd ed. London: T. and J. Allman. Reprint, 1973, Kraus Reprint Co.

Elking, H. 1859. "A View of the Greenland Trade and Whale Fishery, with the National and Private Advantages Thereof." In *A Select Collection of Scarce and Valuable Tracks on Commerce*, edited by J.R. McCulloch, 63–103. London: Harrison and Sons.

Elliott, G. 1998. *A Whaling Enterprise: Salvesen in the Antarctic*. Norwich, U.K.: Michael Russell.

Ellis, R. 1980. *The Book of Whales*. New York: Alfred A. Knopf.

– 1992. *Men and Whales*. London: Robert Hale Ltd.

"Emergence of a Newfoundland Society." n.d. NLHWS.

Enclyclopedia of Newfoundland and Labrador. 1981–1994. Edited by J.R. Smallwood and C.F. Poole. Vols. 1–5. St John's: Newfoundland Book Publishers and Harry Cuff Publications.

England, G.A. 1969. *The Greatest Hunt in the World*. Montreal: Tundra Books (originally published in 1924 as *Vikings of the Ice*).

English, C. 1990. "The Development of the Newfoundland Legal System to 1815." *Acadienis* 20: 89–119.

"English Fishery and Trade in the Eighteenth Century." 1997. NLHWS.

"English Migratory Fishery and Trade in the Seventeenth Century." 1997. NLHWS.

"English Voyages before Cabot." 1997. NLHWS.

Falnes, J. 1999. *Løst Krutt og Knallpester: Hvalfangstbestyren som ble fabrikkeier i Stokke*. Sandefjord: Chr. Christensens Hvalfangstmuseum.

Fay, C.R. 1956. *Life and Labour in Newfoundland*. Cambridge: University of Cambridge.

Field, A.M. 1924. "The Development of Government in Newfoundland, 1638–1713." Master's thesis, University of London.

"Fisheries." 1998. NLHWS.

Fitzgerald, J. 1997. "Conflict and Culture in Irish-Newfoundland Roman Catholicism, 1829–1850." PhD thesis, University of Ottawa.

Flynn, D. 2002. "Scratching the Surface: Story of the Conception Harbour Ship Wrecks and the Last Whaler." *Compass* (Conception Bay) Three-part series, 24 September, 1, 8 October.

Francis, D. 1990. A History of World Whaling. Markham, Ont.: Viking Press.

Fraser, F.C. 1937. "Early Japanese Whaling." *Proceedings of the Linnean Society of London*: 19–21.

Gad, F. 1971. *The History of Greenland*. Vol. 1. Montreal: McGill-Queen's University Press.

Gaskin, D.E. 1972. *Whales, Dolphins and Seals*. London: Heinemann Educational Books.

Gerritsz, H. 1904. "History of the Country called Spitsbergen, 1613." In *Early Dutch and English Voyages to Spitsbergen in the Seventeenth Century*, edited by M. Conway, 9–38. London: Hakluyt Society.

Gilchrist, J. 1972. "Exploration and Enterprise – The Newfoundland Fishery c. 1497–1677." In *Canadian Business History: Selected Studies*, 1497–1971, edited by D. MacMillan, 7–26. Toronto: McClelland and Stewart.

Gillies, R.P. 1826. *Tales of a Voyager to the Arctic Oceans*. Vols. 1–3. London: Henry Colburn.

Goddard, J. 1986. "Reminiscences of Life on the Shore Whaling Stations of British Columbia, 1905–1918." In *Archaeology in Solution: Proceedings of the Seventeenth Annual Conference on Underwater Archaeology*, edited by J.W. Foster and S.O. Smith, 205–11. Sacramento, Calf.: Coyote Press.

Gosling, W.G. 1910. *Labrador: Its Discovery, Exploration and Development*. London: Alston Rivers.

Gosse, E. 1890. *The Life of Philip Henry Gosse*. London: Kegan, Paul, Trench, Trubner and Co.

"Government and Politics." n.d. NLHWS.

Greene, J.P. 1970. "The Influence of Religion in the Politics of Newfoundland." Master's thesis, Memorial University of Newfoundland.

Greene, W.H. 1933. *The Wooden Walls among the Ice Floes: Telling the Romance of the Newfoundland Seal Fishery*. London: Hutchinson and Co.

Grenfell, W.T. 1895. *Vikings of Today, or, Life and Medical Work among the Fishermen of Labrador*. London: Marshall Bros.

– 1906. *Northern Neighbours: Stories of the Labrador People.* London: Hodder and Stoughton.
– 1909. *Labrador: The Country and the People.* New York: Macmillan.
– 1910. *Down to the Sea: Yarns from the Labrador.* New York: Fleming H. Revell.
– 1912. *Down North on the Labrador.* London: James Nisbet.
– 1916. *Tales of Labrador.* London: James Nisbet.
– 1919. *A Labrador Doctor: The Autobiography of Wilfred Thomason Grenfell, M.D. (Oxon), C.M.G.* New York: Houghton Mifflin.
Grenier, R. 1985. "Excavating a 400-Year-Old Basque Galleon." *National Geographic* 168, no. 1: 58–67.
Gulland, J.E., and L.K. Boerema. 1973. "Scientific Advice on Catch Levels." *Report of the International Whaling Commission* 23: 200–12.
Gunn, G.E. 1977. *The Political History of Newfoundland, 1832–1864.* Toronto: University of Toronto Press.
Hagelund, W.A. 1982. *Whalers No More: A History of Whaling on the West Coast.* Madeira Park, B.C.: Harbour Publishing.
Hakluyt, R. 1965. *The Principal Navigations, Voyages, Traffiques and Discoveries of the English Nation.* Vols. 1–12. New York: AMS Press Inc. Reprint of 2nd ed., vols. 1 (1598), 2 (1599), and 3 (1600). London: George Bishop, Ralph Newberrie, and Robert Barker.
Handcock, W.G. 1977. "English Migration to Newfoundland." In *The Peopling of Newfoundland: Essays in Historical Geography*, edited by J.J. Mannion, 15–48. St John's: ISER Press.
– 1979. "An Historical Geography of English Settlement in Newfoundland: A Study of the Migration Process." PhD thesis, University of Birmingham.
– 1981. "The West Country Migrations to Newfoundland." *Bulletin of Canadian Studies* 5, no. 1: 5–24.
– 1985. "The Poole Mercantile Development and Growth of Trinity, 1700–1839." *Newfoundland Quarterly* 80, no. 3: 19–30.
– 1989. *Soe Longe As There Comes Noe Women: Origins of English Settlement in Newfoundland.* St John's: Breakwater Press.
– 1994. "Settlement." In *ENL,* vol. 5, 133–42.
Handcock, W.G. and A. Macpherson. 1987. "Trinity, Eighteenth Century." *Historical Atlas of Canada*, vol. 1., edited by R.C. Harris and G.J. Matthews, pl. 26. Toronto: University of Toronto Press.
Hannesson, H.W. 2003. "A Kinder, Gentler Whaler." Letter to the editor from ambassador of Iceland to Canada. *MacLean's* 116, no. 37: 10.
Harmer, S.F. 1928. "History of Whaling." *Proceedings of the Linnean Society of London*, 140th Session: 51–95.
Harrington, M.F. 1986. *Goin' to the Ice: Offbeat History of the Newfoundland Seal Fishery.* St John's: Breakwater Press.
– 1987a. "Offbeat History." *Evening Telegram* (St John's) 19 October, 16 November.
– 1987b. "Bankers, Whalers and Heritage." *Evening Telegram* (St John's) 1 June.
– 1994. "Shipbuilding." In *ENL* vol. 5, 165–73.
Harris, L. 1959. "The First Nine Years of Representative Government." Master's thesis, Memorial University of Newfoundland.
Harris, M. 1992. *Rare Ambition: The Crosbies of Newfoundland.* Toronto: Viking Press.
Harris, R.C., and G.J. Matthews, eds. 1987. *Historical Atlas of Canada,* vol. 1, *From the Beginning to 1800.* Toronto: University of Toronto Press.
Harrison, R.J., and King, J.E. 1965. *Marine Mammals.* London: Hutchinson University Library.
Harrisse, H. 1892. *The Discovery of North America.* London: H. Stevens.

– 1896. *John Cabot, the Discoverer of North America and Sebastian His Son*. London: B.S. Stevens.

Hart, I.B. 2001. *Pesca: The History of Compañia Argentina de Pesca Sociedad Anonima of Bueno Saires: An Account of the Pioneer Modern Whaling and Sealing Company in the Antarctic*. Salcombe, U.K.: Aidan Ellis.

Harvey, M. 1899. *Newfoundland As It Is in 1899: A Handbook and Tourists' Guide*. London: Kegan, Paul, Trench, Trebner and Co.

– 1902. *Newfoundland at the Beginning of the Twentieth Century*. New York: The South Publishing Co.

Hatton, J., and M. Harvey. 1883. *Newfoundland: The Oldest British Colony, Its History, Its Present Condition, and Its Prospects in the Future*. Boston: Doyle and Whittle.

Hay, K.A. 1979. *Guide to Identification of Whales in Newfoundland-Labrador Waters*. St John's: Department of Fisheries and Oceans, Government of Canada.

– 1982. "Aerial Line-Transect Estimates of Abundance of Humpback, Fin, and Long-Finned Pilot-Whales in the Newfoundland-Labrador Area." *Report of the International Whaling Commission* 32: 475–86.

– 1985. "Status of the Humpback Whale, *Megaptera novaeangliae*, in Canada." *Canadian Field-Naturalist* 99, no. 3: 425–32.

Head, C.G. 1976. *Eighteenth Century Newfoundland: A Geographer's Perspective*. Toronto: McClelland and Stewart.

Head, C.G., C. More, and M. Barkham. 1987. "The Fishery in Atlantic Commerce." *Historical Atlas of Canada*, vol. 1, edited by R.C. Harris and G.J. Matthews, pl. 28. Toronto: University of Toronto Press.

Headland, R.K. 1984. *The Island of South Georgia*. Cambridge: University of Cambridge.

– 1987. "Hvalfangernes Gravsteder i Antarktis." Sandefjord: Chr. Christensens Hvalfangstmuseum 1981–86 (Year Book): 9–38.

Hershkovitz, P. 1966. *Catalogue of Living Whales*. Smithsonian Institution Bulletin 246. Washington.

Hiller, J.K. 1980. "The Railway and Local Politics in Newfoundland, 1870–1901." In *Newfoundland in the Nineteenth and Twentieth Centuries: Essays in Interpretation*, edited by J.K. Hiller and P. Neary, 123–47. Toronto: University of Toronto Press.

– 1990. "The Newfoundland Credit System: An Interpretation." In *Merchant Credit and Labour Strategies in Historical Perspectives*, edited by R.E. Ommer, 90–108. Fredericton, N.B.: Acadiensis Press.

– 1997. "Sponsored Settlement: The Colonization of Newfoundland." NLHWS.

– 2001a. "Sailing Seal Fishery." NLHWS.

– 2001b. "The Nineteenth Century Steamer Seal Fishery." NLHWS.

– 2001c. "Twentieth Century Steamer Seal Fishery." NLHWS.

Hiller, J.K., and P. Neary, eds. 1980. *Newfoundland in the Nineteenth and Twentieth Centuries: Essays in Interpretation*. Toronto: University of Toronto Press.

– 1993. *Twentieth-Century Newfoundland: Explorations*. St John's: Breakwater Press.

Hodgson, P.M. 1981. "Exports." In *ENL*, vol. 1, 793–7.

Hohman, E.P. 1928. *The American Whaleman*. Toronto: Longmans, Green and Co.

Holt, S.J. 2002. "The Whaling Controversy." *Fisheries Research* 54, no. 14: 5–151.

House, D. 1985. *The Challenge of Oil: Newfoundland's Quest for Controlled Development*. St John's: Institute of Social and Economic Research, Memorial University of Newfoundland.

– 1986. *Fish vs. Oil: Resources and Rural Development in North Atlantic Societies*. St John's: Institute of Social and Economic Research, Memorial University of Newfoundland.

– 1988. "Canadian Fisheries Policies and Troubled Newfoundland Communities." In *A Question of Survival: The Fisheries and Newfoundland Society*, edited by P.R. Sinclair,

178–202. St John's: Institute of Social and Economic Research, Memorial University of Newfoundland.

– 1998a. "Industrialization and Diversification." NLHWS.

– 1998b. "Strategies of Diversification." NLHWS.

Howley, J.P. 1915. *The Beothuks, or Red Indians: The Aboriginal Inhabitants of Newfoundland.* Cambridge: Cambridge University Press.

Huswick, A. 1912. "The War on the Whale: Part 2 – The Business End of Whaling." *British Columbia Magazine* 8, no. 1: 7–12.

"Hvalfangernes Gravsteder i Antarkis." 1981–86. *Sandefjordmucene Arbok*: 9–38.

"Icelandic Whale Hunt Underway." *St John's Telegram*, 18 August 2003.

Ingstad, A.S. 1977. *The Discovery of a Norse Settlement in America: Excavations of L'Anse aux Meadows, Newfoundland, 1961–1968.* Vol. 1. Oslo: Universitetsforlaget.

Ingstad, A.S., and H.M. Ingstad. 1986. *The Norse Discovery of North America.* 2 vols. Oslo: Universitetsforlaget.

Ingstad, H.M. 1969. *Westward to Vinland: The Discovery of Pre-Columbian Norse House-Sites in North America.* Toronto: Macmillan.

Innis, H.A. 1954. *The Cod Fisheries: The History of an International Economy.* Toronto: University of Toronto Press.

"International Fishery of the Sixteenth Century." 1997. NLHWS.

"Irish Monks and the Voyages of St. Brendan." 1997. NLHWS.

Iverson, N., and D.R. Matthews. 1968. *Communities in Decline: An Examination of Household Resettlement in Newfoundland.* St John's: Institute of Social and Economic Research, Memorial University of Newfoundland.

Jackson, F.L. 1984. *Newfoundland in Canada: A People in Search of a Polity.* St John's: Harry Cuff Press.

Jackson, G. 1978. *The British Whaling Trade.* Hamden, Conn.: Archon Books.

– 1981. "The Rise and Fall of English Whaling in the Seventeenth Century." In *Proceedings of the International Symposium: Early European Exploitation of the Northern Atlantic, 800–1700*, 55–68. Groningen, Netherlands: Arctic Centre, University of Groningen.

Janzen, O.U. 2001a. "Immigration." NLHWS.

– 2001b. "Settlement and Society." NLHWS.

– 2003. "A Readers' Guide to Newfoundland History to 1869." NLHWS.

Jenkins, J.T. 1921. *A History of the Whale Fisheries: From the Basque Fisheries of the Tenth Century to the Hunting of the Finner Whale at the Present Date.* Reprint, 1971, London: Kennikat Press.

Johnsen, A.O. 1947. *Norwegian Patents Relating to Whaling Industry: A Statistical and Historical Analysis.* Oslo: A.W. Brøggers.

Jones, D. 2000. "R.O. Jones' Letters from the Newfoundland Coast." *Newfoundland Studies* 16, no. 2: 205–26.

Jones, G. 1964. *The Norse Atlantic Saga, Being the Norse Voyages of Discovery and Settlement to Iceland, Greenland, and North America.* Oxford: Oxford University Press.

Journal of the Newfoundland House of Assembly. 1903–30. St John's: Newfoundland Government.

Jukes, J.B. 1842. *Excursions in and about Newfoundland During the Years 1839 and 1840.* Vol. 1. London: John Murray.

Kean, A. 1935. *Old and Young Ahead.* London: Heath Cranton.

Kennedy, J.C. 1991. "Inuit." NLHWS.

Kristjansson, J. 1970. *Icelandic Sagas and Manuscripts.* Reykjevik, Iceland: Saga Publishing Co.

Laing, J. 1822. *A Voyage to Spitsbergen; Containing an Account of That Country; of the Zoology of the North; of the Shetland Islands; and of the Whale Fishery.* Edinburgh: Printed for the Author.

Lang, A. 1908. "Letter to the Editor." *Falkland Islands Magazine and Church Paper* 19, no. 12: 3.

Leslie, J., R. Jameson, and H. Murray. 1831. *Narrative of Discovery and Adventure in the Polar Seas and Regions: With Illustrations of Their Climate, Geology, and Natural History; and an Account of the Whale-Fishery*. 2nd ed. Edinburgh: Oliver and Boyd.

Lien, J. 2001. "The Conservation Basis for the Regulation of Whale Watching in Canada by the Department of Fisheries and Oceans: A Precautionary Approach." *Canadian Technical Report of Fisheries and Aquatic Sciences*, no. 2363: 1–24.

Lien, J., L. Fawcett, and S. Staniford. 1985. *Wet and Fat: Whales and Seals of Newfoundland and Labrador*. St John's: Breakwater Press.

Lounsbury, R.G. 1934. *The British Fishery at Newfoundland, 1634–1763*. New Haven, Conn.: Yale University Press.

Lubbock, B. 1937. *The Arctic Whalers*. Glasgow: Brown, Son and Ferguson.

Lucus, F.A. 1933. *Fifty Years of Museum Work: Autobiography, Unpublished Papers, and Biography of Frederic A. Lucus, Sc.D.* New York: American Museum of Natural History.

McCann, P. 1988. "Culture, State Formation and the Invention of Tradition: Newfoundland, 1832–1855." *Journal of Canadian Studies* 23: 86–103.

McCourt, B. n.d. *The Story of the Whale*: Victoria, B.C.: Pacific Whaling Co.

McCulloch, J.R., ed. 1859. *A Select of Scarce and Valuable Tracts on Commerce*. London: Harrison and Son.

McGhee, R.J. 1996. *Ancient People of the Arctic*. Vancouver: University of British Columbia Press.

McGhee, R.J. and J. Tuck. 1977. "Did the Medieval Irish Visit Newfoundland?" *Canadian Geographical Journal* 94, no. 3: 66–73.

McGrath, P. 1978. "Bristol and America." In *The Westward Enterprise: English Activities in Ireland, the Atlantic, and America, 1480–1650*, edited by K.R. Andrews et al. Liverpool: Liverpool University Press.

McGrath, P.T. 1911. *Newfoundland in 1911*. London: Whitehead, Morris and Co.

MacKintosh, N.A. 1965. *The Stocks of Whales*. London: Fishing News (Books) Ltd.

McLintock, A.H. 1941. *The Establishment of Constitutional Government in Newfoundland, 1783–1832: A Study of Retarded Colonisation*. London: Longmans, Green and Co.

Macpherson, A. 1997. "Pre-Columbian Discoveries and Exploration of North America." In *North American Explorations*, vol. 1, edited by J.L. Allen, 13–70. Lincoln, Neb.: University of Nebraska Press.

Macpherson, A., and B. Wallace. 1987. "Norse Voyages and Settlement." In *Historical Atlas of Canada*, vol. 1, edited by R.C. Harris and G.J. Matthews, pl. 16. Toronto: University of Toronto Press.

Macy, O. 1835. *The History of Nantucket: Being a Compendious Account of the First Settlement of the Island by the English, together with the Rise and Progress of the Whale Fishery*. Boston: Hilliard, Gray, and Co. Reprint, 1970 by Research Reprints Inc., New York.

Madden, D.C. 1981. "Exploration and Discovery." In ENL, vol. 1, 788–92.

Magnusson, M., and H. Palsson (trans.). 1980. *The Vinland Sagas: The Norse Discovery of North America*. London: Penguin Books.

Mannion, J.J. 1974. *Irish Settlements in Eastern Canada: A Study in Cultural Transfer and Adaptation*. Toronto: University of Toronto Press.

– 1976. *Point Lance in Transition: The Transformation of a Newfoundland Outport*. Toronto: McClelland and Stewart.

– 1982. "The Waterford Merchants and the Irish-Newfoundland Provisions Trade, 1770–1820." In *Canadian Papers in Rural History* 3, edited by D.H. Akenson, 178–203. Gananoque: Langdale Press.

– 1986. "Irish Merchants Abroad: The Newfoundland Experience, 1750–1850." *Newfoundland Studies* 2, no. 2: 127–90.

– 1987a. "A Transatlantic Merchant Fishery: Richard Welsh of New Ross and the Sweetmans of Newbawn in Newfoundland, 1734–1862." In *Wexford: History and Society: Interdisciplinary Essays on the History of an Irish County*, edited by K. Whalen and W. Noland, 373–421, 543–5. Dublin: Geography Publications.
– 1987b. "St John's Town, 1728." In *Historical Atlas of Canada*, vol. 1, edited by R.C. Harris and G.J. Matthews, pl. 27. Toronto: University of Toronto Press.
– 1988. "The Maritime Trade of Waterford in the Eighteenth Century." In *Common Ground: Essays on the Historical Geography of Ireland*, edited by W. Smyth and K. Whalen, 208–33. Cork: Cork University Press.
– 1989. "Old World Antecedents, New World Adaptations: Inistioge (Co. Kilkenny) Immigrants in Newfoundland." *Newfoundland Studies* 5, no. 2: 103–76.
– 2001. "Irish Migration and Settlement in Newfoundland: The Formative Phase, 1697–1732." *Newfoundland Studies* 17, no. 2: 257–93.
Mannion, J.J., and C.G. Head. 1987. "The Migratory Fisheries." In *Historical Atlas of Canada*, vol. 1, edited by R.C. Harris and G.J. Matthews, pl. 21. Toronto: University of Toronto Press.
Mannion, J.J., and S. Barkham. 1987. "The 16th Century Fishery." In ibid., pl. 22.
Mannion, J.J., and W.G. Handcock. 1987. "The 17th Century Fishery." In ibid., pl. 33.
Mannion, J.J., W.G. Handcock, and A.G. Macpherson. 1987. "The Newfoundland Fishery, 18th Century." In ibid., pl. 25.
Mansfield, A.W. 1985. "Status of the Blue Whale, *Balaenoptera musculus*, in Canada." *Canadian Field-Naturalist* 99, no 3: 417–20.
Marcus, G.J. 1980. *The Conquest of the North Atlantic*. Woodbridge, Suffolk: Boydell Press.
Markham, M. 1881. "On the Whale-Fishery of the Basque Provinces of Spain." *Proceedings of the Zoological Society of London*, 13 December 1881: 969–76.
Marshall, I. 1996. *A History and Ethnography of the Beothuk*. Montreal: McGill-Queen's University Press.
Martens, F. 1855. "Voyage to Spitsbergen." In *A Collection of Documents on Spitzbergen and Greenland*, edited by A. White, 3–174. London: Hakluyt Society.
Mathieson, J. 1934. "The Study of Arctic Voyages and Exploration." *Royal Scottish Geographical Society Magazine* 50, no. 5: 281–308.
Matthews, K. 1968. "A History of the West of England-Newfoundland Fishery." PhD diss., Oxford University.
– 1975. *Collection and Commentary on the Constitutional Laws of Seventeenth Century Newfoundland*. St John's: Maritime History Group, Memorial University of Newfoundland.
– 1978. "Historical Fence Building: A Critique of the Historiography of Newfoundland." *Newfoundland Quarterly* 74, no. 1: 21–9.
Matthews, L.H., ed. 1968. *The Whale*. London: George Allen and Unwin Ltd.
– 1978. *The Natural History of the Whale*. London: Weidenfield and Nicolson.
Mayo, H.B. 1948. "Newfoundland and Confederation in the Eighteen Sixties." *Canadian Historical Review* 29, no. 2: 125–42.
Mercantile List and Maritime Directory. 1901. St John's: Maritime History Archive, Memorial University of Newfoundland.
– 1904. St John's: Maritime History Archive, Memorial University of Newfoundland.
– 1907. St John's. Maritime History Archive, Memorial University of Newfoundland.
Mercer, M.C. 1975. "Modified Leslie-DeLury Population Models of the Long-Finned Pilot Whale (*Globicephala melaena*) and Annual Production of the Short-Finned Squid (Ilex illecebrosus) Based Upon their Interaction at Newfoundland." *Journal of the Fisheries Research Board of Canada* 32, no. 7: 1145–54.

Meredith, G.R., and R.R. Campbell. 1988. "Status of the Fin Whale, *Balaenoptera physalus*, in Canada." *Canadian Field-Naturalist* 102, no. 2: 351–68.

Millais, J.G. 1907. *Newfoundland and Its Untrodden Ways*. London: Longmans, Green and Co.

Miller-Pitt, J.E. 1981. "Economy." In *ENL*, vol. 1, 667–75.

Miller-Pitt, J.E. and L.E. Russell. 1988. "France." In *ENL*, vol. 2, 369–89.

Mitchell, E. 1972a. "Memorandum on Northwest Atlantic Sei Whales." *Report of the International Whaling Commission* 22: 119–20.

– 1972b. "Assessments of Northwest Atlantic Fin Whale Stocks." *Report of the International Whaling Commission* 22: 111–18.

– 1973. "Draft Report on Humpback Whales Taken under Special Permit by Eastern Canadian Land Stations, 1969–71." *Report of the International Whaling Commission* 23: 138–54.

– 1974a. "Present Status of Northeast Atlantic Fin and Other Whale Stocks." In *The Whale Problem: A Status Report*, edited by W.E. Schevill, 108–69. Cambridge, Mass.: Harvard University Press.

– 1974b. "Canada. Progress Report on Whale Research, May 1972 to May 1973." *Report of the International Whaling Commission* 24: 196–210.

– 1975a. "Review of Biology and Fisheries for Small Cetaceans." *Journal of the Fisheries Research Board of Canada* 32, no. 7: 985–94.

– 1975b. "Preliminary Report of the Nova Scotia Fishery for Sei Whales, *Balaenoptera borealis*." *Report of the International Whaling Commission* 25: 218–24.

Mitchell, E., and R.R. Reeves. 1981. "Catch History and Cumulative Catch Estimates of Initial Population Size of Cetaceans in the Eastern Canadian Arctic." *Report of the International Whaling Commission* 31: 645–82.

– 1983. "Catch History, Abundance and Present Status of Northwest Atlantic Humpback Whales." In *Historical Whaling Records*, edited by M.F. Tillman and G.P. Donovan, 153–212. Cambridge, U.K.: IWC.

Mitchell, E., and V.M. Kozicki. 1975. "Supplementary Information on Minke Whales (*Balaenoptera acutorarostrata*) from the Newfoundland Fishery." *Journal of the Fisheries Research Board of Canada* 32, no. 7: 985–94.

Moores, B. 1993. "Pulp and Paper Manufacture." In *ENL*, vol. 4, 471–8.

– 1994. "Sawmilling." In *ENL*, vol. 5, 92–6.

Morison, S.E. 1971. *The European Discovery of America: The Northern Voyages, 500–1600 AD*. New York: Oxford University Press.

Morris, P. 1827. *Remarks on the State of Society, Religion, Morals, and Education in Newfoundland*. London: A. Hancock.

– 1848. *A Short Review of the History, Government, Constitution, Fishery and Agriculture of Newfoundland*. St John's: J. Woods.

Moyles, R.G. 1975. *"Complaints Is Many and Various, but the Odd Devil Likes It": Nineteenth Century Views of Newfoundland*. Toronto: Peter Martin Associates.

Murphy, R.C. 1929. "Frederic August Lucas: 1852–1929." *Natural History* 29, no. 2: 211.

Murphy, T,. and C.J. Byrne, eds. 1987. *Religion and Identity: The Experience of Irish and Scottish Catholics in Atlantic Canada*. St John's: Jesperson Press.

Murray, J., ed. 1968. *The Newfoundland Journal of Aaron Thomas, 1794*. Don Mills, Ont.: Longmans.

Neary, P. 1978. "Newfoundland and Quebec: Provincial Neighbours across an Uneasy Frontier." *Bulletin of Canadian Studies* 2, no. 2: 35–51.

– 1982. "Canadian Immigration Policy and the Newfoundlanders, 1912–1939." *Acadiensis* 11, no. 2: 69–73.

– 1988. *Newfoundland in the North Atlantic World, 1929–1949*. Montreal: McGill-Queen's University Press.

– 1997. *White Tie and Decorations: Sir John and Lady Hope Simpson in Newfoundland, 1934–1936*. Toronto: University of Toronto Press.

Neary, P., and P. O'Flaherty. 1974. *By Great Waters*. Toronto: University of Toronto Press.

Neary, S. 1994. *The Enemy on Our Doorstep: The German Attacks at Bell Island, 1942*. St John's: Jesperson Press.

Nelson, D., and J. Lien 1996. "The Status of the Long-Finned Pilot Whale, *Globicephala melas*, in Canada." *Canadian Field-Naturalist* 110, no. 3: 521–4.

Nemec, T.F. 1972. "The Irish Emigrations to Newfoundland." *Newfoundland Quarterly* 69, no. 1: 15–19, 22–4.

– 1991. "Irish Settlement in Newfoundland." In *ENL*, vol. 3, 71–7.

"Newfoundland Settlement and the Migratory Fishery." n.d. NLHWS.

Newfoundland. *Annual [Fisheries] Report for 1897*. 1898. St John's: Department of Fisheries, Government of Newfoundland.

– 1902. 1903. St John's: Department of Fisheries, Government of Newfoundland.

– 1903. 1904. St John's: Department of Fisheries, Government of Newfoundland.

– 1904. 1905. St John's: Department of Fisheries, Government of Newfoundland.

– 1905. 1906. St John's: Department of Fisheries, Government of Newfoundland.

– 1906. 1907. St John's: Department of Fisheries, Government of Newfoundland.

– 1907. 1908. St John's: Department of Fisheries, Government of Newfoundland.

Niemi, E. 1993. "Modern Whaling on the Norwegian Arctic Coast – Origin, Development and the Local Society." In *Whaling and History: Perspectives on the Evolution of the Industry*, edited by B.L. Basberg et al., 67–79. Sandefjord: Chr. Christensens Hvalfangstmuseum.

"Norse in the North Atlantic." 1997. NLHWS.

O'Dea, F. 1988. "Cabot's Landfall – Yet Again." In *Explorations in the History of Canadian Mapping: A Collection of Essays*, edited by B. Farrell and A. Desbarats. Ottawa: Association of Canadian Map Libraries and Archives.

O'Flaherty, P. 1979. *The Rock Observed: Studies in the Literature of Newfoundland*. Toronto: University of Toronto Press.

– 1999. *Olde Newfoundland: A History to 1843*. St John's: Long Beach Press.

Ommanney, F.D. 1971. *Lost Leviathan*. London: Hutchinson & Co.

Ommer, R.E. 1983. "'A Peculiar and Immediate Dependence of the Crown': The Basis of the Jersey Merchant Triangle." *Business History* 25, no. 2: 107–24.

– 1987. "The Cod Trade in the New World." In *A People of the Sea: The Merchant History of the Channel Islands*, edited by A. Jamieson, 245–68. London: Methuen.

– 1990. *Merchant Credit and Labour Strategies in Historical Perspective*. Fredericton, N.B.: Acadiensis Press.

– 1991. *From Outport to Outport: A Structural Analysis of the Jersey-Gaspe Cod Fishery, 1767–1886*. Montreal: McGill-Queen's University Press.

O'Neill, P. 1975, 1976. *The Oldest City: The Story of St John's, Newfoundland*. 2 vols. Erin, Ont.: Porcepic Press.

"Organizing the Migratory Fishery." 1997. NLHWS.

"Other Claims for Pre-Cabot Discovery." 1997. NLHWS.

"Out for Whales." 2003. *Toronto Globe & Mail*, 25 August.

"Palaeo-Eskimo Tradition." 1993. In *ENL*, vol. 4, 200–2.

Pastore, R. 1992. *Shanawdithit's People*. St John's: Atlantic Archaeology.

– 1993. "Archaeology, History and the Beothuks." *Newfoundland Studies* 9, no. 2: 260–77.

– 1994. "The Sixteenth Century: Aboriginal Peoples and European Contact." In *The Atlantic Region to Confederation: A History*, edited by P.A. Buckner and J.G. Reid. Toronto: University of Toronto Press.

– 1998a. "Palaeo-Eskimo Peoples." NLHWS.

– 1998b. "The Thule." NLHWS.

Pedley, C. 1863. *The History of Newfoundland: From the Earliest Times to the Year 1860*. London: Longman, Green, Longman, Roberts, and Green.

Petursddottir, G., ed. 1997. *Conference Proceedings: Whaling in the North Atlantic, Economic and Political Perspectives*. Reykjavik: University of Iceland.

Phillips, R. 1819. *Journal of a Voyage of Discovery, to the Arctic Regions, Performed between the 4th of April and the 18th of December, 1818, in His Majesty's Ship Alexander, Wm. Edw. Perry, Esq. Lieut. and Commander*. London: Printed for the author by G. Sidney.

Pierson, S. 1992. "A Review: *The Afterlife of George Cartwright*, J. Steffer (1922)." *Newfoundland Studies* 8, no. 2: 202–20.

Pope, 1997a. *The Many Landfalls of John Cabot*. Toronto: University of Toronto Press.

– 1997b. "The Sixteenth Century Fishing Voyage." In *How Deep Is the Ocean? Historical Essays on Canada's Atlantic Fishery*, edited by J.E. Candow and C. Corbin, 15–30. Sydney, N.S.: University College of Cape Breton Press.

Powys, L. 1927. *Henry Hudson*. London: Bodley Head.

"Prehistory." 1987. *Historical Atlas of Canada*, vol. 1, edited by R.C. Harris and G.J. Matthews, 1–6. Toronto: University of Toronto Press.

Proulx, J.-P. 1979. *The Military History of Placentia: A Study of French Fortificiations*. History and Archaeology Series, no. 26. Ottawa: Parks Canada.

Prowse, D.W. 1895. *A History of Newfoundland: From the English, Colonial, and Foreign Records*. London: MacMillan and Co. Facsimile edition, 1972, Belleville, Ontario: Mika Studio.

Purchas, S. 1906. *Purchas: His Pilgrimes*. Vols. 1–20. Glasgow: James MacLehose and Sons. First published 1625.

Quinn, D.B. 1974. *England and the Discovery of North America*. New York: Knopf.

– 1977. *North America from Earliest Discovery to First Settlements: The Norse Voyage to 1612*. New York: Harper and Row.

– 1979a. *New American World: A Documentary History of North America to 1612*. New York: Arno Press.

– 1979b. *Newfoundland from Fishery to Colony*. New York: Arno Press.

– 1982. "Newfoundland in the Consciousness of Europe in the Sixteenth and Early Seventeenth Centuries." In *Early European Settlement and Exploration in Atlantic Canada: Selected Papers*, edited by G.M. Storey. St John's: Memorial University of Newfoundland.

– 1993. *Sebastian Cabot and Bristol Exploration*. Bristol: Bristol Branch of the Historical Association. Rev. ed.

Reeves, J. 1793. *History of the Government of the Island of Newfoundland*. Reprint, 1967. New York: Johnson Reprint Corp.

Reeves, R.R., and T.D. Smith. 2002. "Historical Catches of Humpback Whales in the North Atlantic Ocean: An Overview of Sources." *Journal of Cetacean Resource Management* 4, no. 3: 219–34.

– 2003. *A Taxonomy of World Whaling: Operations, Eras, and Data Sources*. Woods Hole, Mass.: Northeast Fisheries Science Reference Document 03–12.

Rennie, R. 1998. "Mining." NLHWS.

Renouf, M.A.P. 1993. "Palaeoeskimo Seal Hunters at Port au Choix, Northwestern Newfoundland." *Newfoundland Studies* 9, no. 2: 185–212.

– 1994. "Two Traditional Sites in Port au Choix, Northwestern Newfoundland." In *Trends of Arctic Prehistory: Papers in Honour of William E. Taylor, Jr*, edited by D. Morrison and J-L. Pilon, 145–64. Mercury Series, no. 149. Hull, Quebec: Canadian Museum of Civilization.

– 1999. *Ancient Cultures, Bountiful Seas: A Story of Port au Choix*. St John's: Historic Sites Association of Newfoundland and Labrador.

Rissmüller, R.H. 1873. "Uber die Stoffwandering in der Pflanze." PhD diss., University of Gottingen (Courtesy of K. Barthelmess).

Robbins, D.T. 1986. "Newfoundland Dorset Culture?" In *Palaeo-Eskimo Cultures in Newfoundland, Labrador and Ungava*, 119–24. Vol. 1, *Reports in Archaeology*. St John's: Memorial University of Newfoundland.

Rompkey, R. 1991. *Grenfell of Labrador: A Biography*. Toronto: University of Toronto Press.

– 1996. *Labrador Odyssey: The Journal and Photographs of Eliot Curwen on the Second Voyage of Wilfred Grenfell, 1893*. Montreal: McGill-Queen's University Press.

Rowe, F.W. 1964. *The Development of Education in Newfoundland*. Toronto: University of Toronto Press.

Rowen, H.H., ed. 1972. *The Low Countries in Early Modern Times*. New York: Harper and Row.

Ruggles, R.I. 1987. "Exploring the Atlantic Coast." In *Historical Atlas of Canada*, vol. 1, edited by R.C. Harris and G.J. Matthews, pl. 19. Toronto: University of Toronto Press.

Ryan, S. 1971. "The Newfoundland Cod Fishery in the Nineteenth Century." Master's thesis, Memorial University of Newfoundland.

– 1980. "The Newfoundland Salt Cod Trade in the Nineteenth Century." In *Newfoundland in the Nineteenth and Twentieth Centuries: Essays in Interpretation*, edited by J.K. Hiller and P. Neary, 40–66. Toronto: University of Toronto Press.

– 1983a. *Newfoundland – Spanish Saltfish Trade: 1814–1914*. St John's: Harry Cuff Publications.

– 1983b. "Fishery to Colony: A Newfoundland Watershed, 1793–1813." *Acadiensis* 11, no. 2: 34–52.

– 1986. *Fish out of Water: The Newfoundland Saltfish Trade, 1814–1914*. St John's: Breakwater Press.

– 1994a. "Sealing." In *ENL*, vol. 5, 118–25.

– 1994b. *The Ice Hunters: A History of Newfoundland Sealing to 1914*. St John's: Breakwater Press.

Salvesen, T.E. 1914. "The Whale Fisheries of the Falkland Islands and Dependencies." *Report of the Scientific Results of the Scottish National Antarctic Expedition*. Edinburgh: Scottish Oceanography Laboratory.

Sanger, C.W. 1973. "Technological and Spatial Adaptation in the Newfoundland Seal Fishery During the Nineteenth Century." Master's thesis, Memorial University of Newfoundland.

– 1977. "The Evolution of Sealing and the Spread of Settlement in Northeastern Newfoundland." In *The Peopling of Newfoundland: Essays in Historical Geography*, edited by J.J. Mannion, 136–51. St John's: ISER Press.

– 1980. "The 19th Century Newfoundland Seal Fishery and the Influence of Scottish Whalemen." *Polar Record* 20, no. 126: 231–52.

– 1985. "The Origins of the Scottish Northern Whale Fishery." PhD diss., University of Dundee.

– 1988. "The Dundee–St John's Connection: Nineteenth Century Interlinkages Between Scottish Arctic Whaling and the Newfoundland Seal Fishery." *Newfoundland Studies* 4, no. 1: 1–26.

– 1991a. "Environmental Factors Affecting Seventeenth-Nineteenth Century Whaling in the Greenland Sea." *Polar Record* 27, no. 161: 77–86.

– 1991b. "'On Good Fishing Ground, but Too Early for Whales I Think': The Impact of Greenland Right Whale Migration Patterns on Hunting Strategies in the Northern Whale Fishery, 1600–1900." *The American Neptune* 51, no. 4: 221–40.

– 1991c. "'Saw Several Finners, But No Whales': The Greenland Right Whale (Bowhead) – An Assessment of the Biological Basis of the Northern Whale Fishery during the Seventeenth, Eighteenth and Nineteenth Centuries." *International Journal of Maritime History* 3, no. 1: 127–54.

– 1991d. "Changing Resources and Hunting Locations of Scottish Whaling-Sealing Vessels in the Second Half of the NineteenthCentury." *Scottish Geographical Magazine* 107, no. 3: 187–97.

– 1994a. "'Dodging in the Bight, A Good Place for a Whale': Environmental Factors Affecting Eighteenth and Nineteenth Century Whaling in Davis Strait." *Northern Mariner/Le Marin du Nord* 4, no. 1: 17–33.

– 1994b. "'We Are Now in a Splendid Position for Whales': Environmental Factors Affecting Nineteenth Century Whaling in Baffin Bay." *Mariner's Mirror* 80, no. 2: 159–77.

– 1998a. "Marine Resources Use." NLHWS.

– 1998b. "Seal Fishery: Background." NLHWS.

– 1998c. "Seal Fishery Equipment Technologies and Products." NLHWS.

– 1998d. "Seal Fishery Hunting Methods." NLHWS.

– 1998e. "Seal Fishery in the Twentieth Century." NLHWS.

Sanger, C.W., and A.B. Dickinson. 1989. "The Origins of Modern Shore Based Whaling in Newfoundland and Labrador: The Cabot Steam Whaling Co. Ltd, 1896–98." *International Journal of Maritime History* 1, no. 1: 129–57.

– 1991a. "Expansion of Regulated Modern Shore-Station Whaling in Newfoundland and Labrador, 1902–3." *Northern Mariner/Le Marin du Nord* 1, no. 2: 1–22.

– 1991b. *"They Were Clannish As Hell": Origins of Shore-Station Whaling in British Columbia – The Newfoundland Factor*. Halifax: Oceans Institute of Canada.

– 1991c. "Newfoundland Involvement in Twentieth Century Shore-Station Whaling in British Columbia." *Newfoundland Studies* 7, no. 2: 97–122.

– 1994. "Whaling." In *ENL* vol. 5, 546–51.

– 1995. "Renewal of Newfoundland and Labrador Shore-Station Whaling, 1918–1936." *International Journal of Maritime History* 7, no. 1: 83–103.

– 1997. "The Construction and Display of the First Full-Scale Model of a Blue Whale: The Newfoundland Connection." *Acadiensis* 27, no. 1: 67–84.

– 1998. "Trinity Involvement in Modern Shore-Station Whaling." *Newfoundland Quarterly* 92, no. 1: 2–6.

– 2000. "Newfoundland and Labrador Shore-Station Whaling: Over Expansion and Decline 1905–1917." *Northern Mariner/Le Marin du Nord* 10, no. 4: 13–38.

Scammon, C.M. 1874. *The Marine Mammals of the North-Western Coast of North America: Together with an Account of the American Whale-Fishery*. San Francisco: John H. Carmany and Co. Reprint, 1968, New York: Dover Publications.

Scaplen, R.P. 1967. "Observations on the Operation of the Whale Catcher *Kyo Maru 17* off Newfoundland." *Progress Report Industrial Development Service*. Ottawa: Department of Fisheries.

Schevill, W.E., ed. 1974. *The Whale Problem: A Status Report*. Cambridge, Mass.: Harvard University Press.

Schmitt, F.P., C. De Jong, and F.H. Winter. 1980. *Thomas Welcome Roys: America's Pioneer of Modern Whaling*. Newport News, Vir.: The Mariners Museum.

Scoresby, W. Jr. 1820. *An Account of the Arctic Regions with a History and Description of the Northern Whale-Fishery*. Vols. 1–2. Edinburgh: Archibald Constable and Co. Reprint, 1969, Newton Abbot, Devon: David and Charles Reprints.

Seitz, D.C. 1926. *The Great Island: Some Observations in and about this Crown Colony of Newfoundland*. New York: The Century Co.

Sergeant, D.E. 1953. "Whaling in Newfoundland and Labrador Waters." *Norsk-Hvlfangst Tidende* 40, no. 12: 687–95.

– 1962. "The Biology of the Pilot or Pothead Whale, *Globicephala melaena* (Traill) in Newfoundland Waters." *Bulletin of the Fisheries Research Board of Canada* 132: 1–84.

– 1963. "Minke Whales, *Balaenoptera acutorostrata* (Lacapede) of the Western North

Atlantic." *Journal of the Fisheries Research Board of Canada* 20, no. 6: 1489–1504.
– 1977. "Stocks of Fin Whales (*Balaenoptera physalus L.*) in the North Atlantic Ocean." *Report of the International Whaling Commission* 27: 357–62.
Sergeant, D.E., and H.D. Fisher. 1957. "The Smaller Cetacea of Eastern Canadian Waters." *Journal of the Fisheries Research Board of Canada* 14, no. 1: 83–115.
Sider, G.S. 1980. "The Ties That Bind: Culture and Agriculture, Property and Propriety in the Newfoundland Village Fishery." *Social History* 5: 1–39.
Smithsonian Institution. 1906. *Annual Report of the Board of Regents (1905)*. Washington, D.C.: Smithsonian Institution.
Sørensen, G. 1912. *Hvalfangsten: Dens Historie Og Maend*. Kristiana: A/S Det Norske Forlagskompani, Ltd.
Southwell, T. 1899. "Notes on the Seal and Whale Fishery, 1899." *Zoologist* 4, no. 38: 108–12.
– 1900. "Notes on the Seal and Whale Fishery, 1900." *Zoologist* 5, no. 51: 81–90.
Starbuck, A. 1878. *History of the American Whale Fishery: From Its Earliest Inception to the Year 1876*. Vols. 1 and 2. Washington, D.C.: Government Printing Office. First published in part 4 of the *Report of the U.S. Commission on Fish and Fisheries*. Washington, D.C. Reprint, 1964, by Argosy-Antiquarian Ltd.
Steel, I. 1986. *The English Atlantic, 1675–1740*. New York: Oxford University Press.
Stephen, Alexander, and Sons. 1932. *A Shipbuilding History, 1750–1932*. London: Printed for Alexander Stephen and Sons Ltd by Ed. J. Burrow and Co.
Stockwood, W. 1991. "Manufacturing and Refining." In *ENL*, vol. 3, 443–9.
Stoett, P.J. 1997. *The International Politics of Whaling*. Vancouver: University of British Columbia Press.
Storey, G.M. ed. 1982. *Early European Settlement and Exploration in Atlantic Canada: Selected Papers*. St John's: Memorial University of Newfoundland.
Talbut, T. 1882. *Newfoundland; or a Letter Addressed to a Friend in Ireland in Relation to the Condition and Circumstances of the Island of Newfoundland, with an Especial View to Emigration*. London: Sampson, Low, Marston, Searle, and Rivington.
Templeman, W. 1966. "Marine Resources of Newfoundland." *Bulletin of the Fisheries Research Board of Canada* 154: 1–170.
Thornton, P.A. 1979. "Dynamic Equilibrium: Settlement, Population and Ecology in the Strait of Belle Isle, Newfoundland." PhD diss., University of Aberdeen.
Tillman, M.F., and G.P. Donovan. 1893. *Historical Whaling Records*. Cambridge, U.K.: IWC.
Tocque,P. 1844. *Wandering Thoughts or Solitary Hours*. London: Thomas Richardson and Son.
– 1895. *Kaleidoscope Echos, Being Historical, Philosophical, Scientific, and Theological Sketches, from Miscellaneous Writings*. Toronto: Hunter, Rose.
Tønnessen, J.N., and Johnsen, A.O. 1982. *The History of Modern Whaling*. Los Angeles: University of California Press. Translated by R.I. Christophersen.
True, F.W. 1904. *The Whalebone Whales of the Western North Atlantic Compared with Those Occurring in European Waters: with Some Observations on the Species of the North Pacific*. Smithsonian Contributions to Knowledge, v. 33. Washington, D.C.: Smithsonian Institution.
– 1903. "On Some Photographs of Living Finback Whales from Newfoundland." *Smithsonian Miscellaneous Collection* 45, no. 1: 91–4.
Tuck, J.A. 1976a. *Newfoundland and Labrador Prehistory*. Ottawa: National Museum of Man.
– 1976b. *Ancient People of Port au Choix*. Newfoundland Social and Economic Studies, no. 17. St John's: Institute of Social and Economic Research, Memorial University of Newfoundland.
– 1985. "Unearthing Red Bay's Whaling History." *National Geographic* 168, no. 1: 50–7.
– 1991. "Maritime Archaic Tradition." In *ENL*, vol. 3, 462–5.
– 1998. "Maritime Archaic Indians." NLHWS.

Tuck, J.A., and R. Grenier. 1981. "A Sixteenth-Century Basque Whaling Station in Labrador." *Scientific American* 245, no. 5: 180–8.

– 1989. *Red Bay; World Whaling Capital A.D. 1550–1600*. St John's: Atlantic Archaeology.

Turgeon, L. 1981. "Pour une histoire de la pêche: le marché de la morue à Marseille au XVIIIe siècle." *Histoire Sociale/Social History* 14, no. 28: 295–322.

Vamplew, W. 1975. *Salvesen of Leith*. Edinburgh: Chr. Salvesen Co.

"Voluntary Settlement: The Peopling of Newfoundland to 1820." NLHWS.

Waring, G.T., J.M. Quintal, and S.L. Swartz, eds. 2001. *U.S. Atlantic and Gulf of Mexico Marine Mammal Stock Assessments, 2000*. NOAA Technical Memorandum, NMFS-NE-168. Woods Hole, Mass.: Northeast Fisheries Science Centre, National Oceanic and Atmospheric Administration.

Watson, L. 1981. *Sea Guide to Whales of the World*. New York: Dutton.

Webb, J.A. 2001a. "Provincial Government: The Smallwood Years, 1949–1972." NLHWS.

– 2001b."Provincial Politics, 1972–2001." NLHWS.

– 2001a."Responsible Government, 1855–1933." NLHWS.

Webb, R.L. 1984. *West Whaling: A Brief History of Whale Hunting in the Pacific Northwest*. Vancouver: Vancouver Whaling Museum.

– 1988a. *On the Northwest: Commercial Whaling in the Pacific Northwest, 1790–1967*. Vancouver: Vancouver Maritime Museum.

– 1988b. *On the Northwest: Commercial Whaling on the Pacific Northwest*. Vancouver: University of British Columbia Press.

Wells, E.A. 1966. "The Struggle for Responsible Government in Newfoundland, 1846–1855." Master's thesis, Memorial University of Newfoundland.

Wexelsen, E. 1993. "Working Conditions and Work-Related Injuries and Illnesses on Board Norwegian Floating Factories after World War II." In *Whaling and History: Perspectives on the Evolution of the Industry*, edited by B.L. Basberg et al., 139–56. Sandefjord: Chr. Christensens Hvalfangstmuseum.

White, A. ed. 1855. *A Collection of Documents on Spitzbergen and Greenland*. London: Hakluyt Society.

White, M. 1987. "A Study of Patrick Morris's Political Pamphlets." Master's thesis, Memorial University of Newfoundland.

Whitehead, H. 1982. "Populations of Humpback Whales in the Northwest Atlantic." *Report of the International Whaling Commission* 32: 345–53.

– 1987. "Updated Status of the Humpback Whale, *Megaptera novaeangliae*, in Canada." *Canadian Field-Naturalist* 101, no. 2: 284–94.

Whiteley, W.H. 1972. "James Cook, Hugh Palliser and the Newfoundland Fishery." *Newfoundland Historical Society* 5, no. 1: 17–22.

Wicks, R. 1974. "Newfoundland Social Life: 1750–1850." *Newfoundland Quarterly* 70, no. 4: 17–23.

Williams, A.F. 1996. *John Cabot and Newfoundland*. St John's: Newfoundland Historical Society.

Willson, G.B. 1897. *The Tenth Island: Being Some Account of Newfoundland, Its People, Its Politics, Its Problems, and Its Peculiarities*. London: Grant Richards.

Wilson, I. 1991. *The Columbus Myth: Did Men of Bristol Reach America before Columbus?* Toronto: Simon and Shuster.

Wilson, W. 1866. *Newfoundland and Its Missionaries*. Halifax: Dakin and Metcalf.

Wix, E. 1836. *Six Months of a Newfoundland Missionary's Journal, from February to August, 1835*. London: Smith, Elder.

Wogn-Henriksen, J.K. 1983–4. "Hvalfangst med rot i Stokke II." In *Stokke: Arv og oett*. Stokke, Norway: Arsskrift for Stokke Historielag.

Wright, J.V. 1987a. "Cultural Sequences, 4000–1000 BC." In *Historical Atlas of Canada*, vol. 1,

edited by R.C. Harris and G. J. Matthews, Toronto, pl. 7. Toronto: University of Toronto Press.
– 1987b. "Cultural Sequences, 1000 BC-AD 500." In ibid., pl. 8.
– 1987c. "Cultural Sequences, AD 500-European Contact." In ibid., pl. 9.
– 1995. *A History of the Native People of Canada.* Vol. 1, Mercury Series, no. 152, Hull, Quebec: Canadian Museum of Civilization.
Wright, J.V., V.K. Prest, and J.-S. Vincent. 1987. "Cultural Sequences, 8000–4000 BC." In *Historical Atlas of Canada*, vol. 1, edited by R.C. Harris and G. J. Matthews, pl. 6. Toronto: University of Toronto Press.
Yonge, J. 1963. *The Journal of James Yonge (1647–1721)*, edited by F.N.L. Payner. London: Longmans Press.

INDEX

COMPANIES

PERSONS

SUBJECTS

VESSELS